LeTourneau Heavy Equipment

A Photographic History from 1921 to Today's Giant Marvels

Eric C. Orlemann

Enthusiast Books
1830A Hanley Road
Hudson, Wisconsin 54016 USA

Enthusiast Books are offered at a discount when sold in quantity for promotional use. Businesses or organizations seeking details should write to the Marketing Department, Enthusiast Books, at the above address.

Library of Congress Control Number: 2014952551

ISBN-13: 978-1-58388-317-4
ISBN-10: 1-58388-317-7

14 15 16 17 18 19 6 5 4 3 2 1

Printed in USA

On the cover:

Front:
The second largest "Electric-Digger" model produced by R.G. LeTourneau was the Series LT-300. Only the LT-360 was bigger. Image date August 1966.

Back:
Pictured on March 7, 1968, is the pilot XL-1 (L-700) "LeTric-Loader." This image was taken behind the main gymnasium on the campus of LeTourneau University, which was located across the street from the main plant in Longview, Texas.

TABLE OF CONTENTS

ACKNOWLEDGMENTS

I would like to thank the many current and retired employees that have worked for LeTourneau and Joy Global that have given me so much help over the years as I try to chronicle the great equipment designs that have carried the "LeTourneau" name over the decades.

I also would like to thank Thomas Berry of the Historical Construction Equipment Association (H.C.E.A.) for the research time he put in for me looking up long forgotten items that needed to be once again brought to everyone's attention. Special thanks also to Keith Haddock of Park Communications, and to Richard L. Smith and Sally Watkins of the Komatsu America Corp. (KAC).

Eric C. Orlemann
Decatur, Illinois
September 2014

The author in the cab of an L-2350 on the Longview, Texas, assembly line taken on April 25, 2007.

INTRODUCTION

If you were to ask someone in North America to name a heavy earth-moving equipment manufacturer, they most likely would come up with the name "Caterpillar." To be fair, if you lived in Japan that name would most likely be "Komatsu," but we are getting off the point! If you pushed a little harder for another name it would probably be "John Deere." But after that, most in the general public would draw a blank. I bet if you could go back a few years and ask the same question, a few more names might come to mind, such as Euclid, Allis Chalmers, Bucyrus Erie, Lima, Lorain, and Marion to name just a few. As corporations merge and gobble-up other companies the classic names of those firms are usually the first to go into the dust-bin of history. This happens in all industries, but it is especially sad to see it happen to firms whose creations literally help build everything we have utilized in the 20th Century and now the 21st. But we as a society have short memories. If we no longer see it, then it must not exist anymore. The names of the equipment we saw growing up working along our highways and streets are gone for the most part, their legacies becoming a dim memory as we age.

For years the name of LeTourneau, while no longer at the forefront in people's memories, still pressed on, creating massive mining machinery out of their Longview, Texas, manufacturing plant. However, it looks as if this historic name will soon be leaving us as well due to the purchase of the company by Joy Global in 2011. As Joy integrates the LeTourneau product lines into its own, the heritage of R. G. LeTourneau's history will start to become blurred as time goes by.

In an effort to keep this haziness at bay, Enthusiast Books and I bring you this book, meant to be a companion publication to my three earlier works concerning LeTourneau published by Enthusiast Books. This Photo History contains many new images that have come about since the printing of the previous volumes, plus many more that simply were edited out due to space limitations. Also, many writing updates and corrections have been made from my previous works and are included here as well in brief introduction and caption form. While the previous books presented the LeTourneau history in three distinct eras (Mechanical Drive, Electric-Drive, and Modern), this volume covers all of the time periods from approximately 1921 to present, told in images not found in the other three works. Along with the required historical background images, extra photographic coverage has been provided for key equipment and machinery designs that have been skipped over in the past. And for all-time favorites like the LT-360 Electric-Digger, extra images have been included to help satisfy the numerous requests I get for these one-of-a-kind marvels.

It is sad to see the LeTourneau name come off the Longview, Texas, manufacturing plant. But Joy Global is investing heavily in Longview, with new paint, brighter working conditions, new machine tools, expanded assembly lines, and increased production due to the shifting of other Joy and P&H work from other plants. I would have to say the future never looked as bright and as secure for the Longview facilities as it does now. I just wish the "LeTourneau" name could have come along for the ride too.

Chapter 1 – The Early Years

R. G. LeTourneau's earliest equipment creations were born out of necessity. Working around the Stockton, California, area as a "land-leveler," to make ends meet and win contracts, he had to offer something his competitors could not. R. G. was always looking for ways to move dirt faster and cheaper. Not satisfied with what the industry was producing equipment-wise to help him meet these needs, he would embark on a journey that would last the rest of his life in the quest of building more productive and innovative earth-moving and material-handling machine creations.

One of the earliest pieces of equipment to be built by LeTourneau was a tractor-towed agricultural ripper (referred to as a subsoiler) in 1920. This was followed by the design and building of his first towed scraper in 1921 known as the "Full-Drag Scraper." After this came the "Gondola" and the "Mountain Mover," both in 1922. All of these early equipment fabrications were utilized by LeTourneau as a land-leveler, and were never built with any future production in mind. They all were a means to an end to make him a more successful contractor. This status as a standalone contractor would start to change with his introduction of the tracked "Earth-Mover" telescopic bucket scraper in 1925. Though the first units built of the tracked-scraper were for his use as a contractor, others in the earth-moving industry would soon take note of his remarkable invention. One person in particular, by the name of Henry J. Kaiser, would change the course of LeTourneau and help him evolve from a contractor to that of a manufacturer. Kaiser offered LeTourneau a six-month employment contract starting in February 1927 in which he would help set up a new manufacturing yard in Livermore, California. Part of this deal also included the sale of LeTourneau's design patents and machine tools from his assembly shop that Kaiser would now control. After the completion of the Livermore facilities, LeTourneau went right to work designing and building equipment for Kaiser, including his telescopic tracked scraper.

After leaving the employment of Kaiser, LeTourneau once again set out on his own. At first he would rent the scrapers he needed from his former employer, but he found the scrapers he could get from Kaiser too slow and unreliable. In 1928 LeTourneau invented a series of equipment designs that were initially utilized by him, and then offered to other earth-moving contractors. The designs included his rear-mounted cable control Power Control Unit (PCU), the 6-cu-yd capacity Cable-controlled Scraper, the 16-cu-yd Dump-Cart, and the Bull-Dozer attachment for track-type tractors. These designs, along with a new 7-cu-yd Highboy scraper, the Sheep's Foot Roller, and the Hard-Pan Rooter, would soon be actively marketed to other contractors in 1929, firmly establishing R. G. LeTourneau as a manufacturer of innovative, heavy-duty earth-moving equipment.

In the summer of 1921, R. G. LeTourneau built his first towed electric-controlled scraper not based on any other existing piece of equipment. Referred to as a "full-drag" scraper, it featured brazed welded construction.

One of LeTourneau's earliest price lists for his new equipment offerings was printed in a small sales catalog in late 1929. Conducting business under the name of "The LeTourneau Manufacturing Co.," the catalog listed the following pieces of equipment and their prices available for purchase:

Not long after the release of this price list, the company was incorporated in November 1929 in the state of California to conduct business under the new name of "R.G. LeTourneau, Inc."

The Power Control Unit, Single $490.00
The Power Control Unit, Double $750.00
Chariot Type Dump Cart, Without Control Unit. $2,500.00
Chariot Type Dump Cart, Body and Frame Only $1,500.00
7-Yard Scraper, Without Control Unit $2,550.00
Hardpan Rooter, Without Control unit $1,185.00
Bull-Dozer, Without Control Unit $985.00
Sheep's-Foot Roller, 2, 4-Foot Sections $925.00

R. G. LeTourneau's "Mountain Mover" scraper was built in the summer of 1922, and featured brazed welded construction, electric motor controls, and a telescoping bucket design. It survives to this day and can been seen on permanent display at the R. G. LeTourneau Memorial Park, located on the grounds of LeTourneau University, Longview, Texas.

The "Self-propelled Scraper," designed and built by LeTourneau in 1923, was the industry's first scraper to move and operate under its own power. An all-electric drive design, powered by a front-mounted gasoline engine and dynamo set, it was capable of carrying a 12-cu-yd payload. But because of its slow speed, it was eventually converted to a tractor-drawn scraper unit and sold around 1925. Image date is 1924.

In 1924 R. G. LeTourneau designed a one-off piece of equipment that he referred to as a "back-filler," for use on a rugged 12-mile stretch of the Hetchy-Hetchy water pipeline project at Redwood City, California. The 50-foot traveling crane was built to straddle the spoil bank and operate at ground level. Electric power for the bucket pulley system was supplied by a generator attached to his Holt 75 tractor. *Image courtesy of KAC.*

R. G. LeTourneau's tracked "Earth-Mover" five-bucket telescoping scraper design was originally introduced in 1925 (also referred to as a Tracked Telescopic Scraper). LeTourneau produced approximately three units before he sold the patents for the design to the Warren Brothers Company (in a deal brokered by Henry J. Kaiser and signed in January 1927). During a six-month employment stint with Kaiser starting in 1927, LeTourneau helped produce more tracked Earth-Movers utilizing hydraulic controls instead of his original electric designs. Kaiser would continue to build the Earth-Mover scrapers after R. G. left his employment, many of which had their tracks removed and replaced with four steel wheels, running two per side. Pictured in 1926 is one of the original three Earth-Mover towed scrapers.

In February 1927, R. G. LeTourneau started a six-month employment contract with Henry J. Kaiser. His first assignment was to build a new equipment facility in Livermore, California, for Kaiser. After that he designed and built a new scraper model referred to as the Kaiser-Le-Tourneau Gondola. Pictured in 1927, in the Kaiser Paving Co. Livermore yard, is the hydraulic controlled Gondola (later converted to electric controls). In the background are at least six more of the hydraulic controlled tracked "Earth-Mover" scrapers ready for shipment. Also of note is the overhead crane in the back of the yard fabricated out of R. G.'s old "back-filler" from 1924. *Image courtesy KAC.*

R. G. LeTourneau originally built an electric controlled towed blade-grader in the summer of 1926 (pictured) for use on an earth-moving contract with Henry J. Kaiser on the Philbrook Dam project, located on the Feather River in Butte County, California. *Image courtesy KAC.*

LeTourneau would improve upon his earlier blade-grader design and build a cable-controlled version of one in 1929 featuring a 13-foot wide moldboard, for use on a Benicia, California earth-moving contract. This blade grader utilized a frame design very similar to the very first "Cable-controlled Scraper" he built for use on the Oroville, California, contract in the fall of 1928.

In 1929 R. G. LeTourneau introduced his 7-cu-yd capacity "Highboy" cable controlled scraper and double-drum Power Control Unit (PCU). Number of Highboy scrapers built was listed at 29. Pictured in 1930 is a Highboy attached to a Caterpillar "Sixty" track-type tractor.

LeTourneau continued to improve his cable-controlled scraper offerings and introduced his 7-cu-yd "Lowboy" in late 1930 as a replacement for the Highboy. The Lowboy, like its predecessor the Highboy, were both considered semi-drag scraper designs. Approximately 42 would be produced. Pictured in March 1933 is a Caterpillar "Sixty" with a Lowboy in tow.

LeTourneau's new type "A" Carryall was originally introduced in 1932. Though normally sold with steel wheels, LeTourneau fitted a set of pneumatic tires to one of his A Carryalls owned by a contractor working in the Imperial Valley of California. Pictured in August 1932 is that Carryall, which is considered the first scraper to be equipped with tires for earth-moving applications, and not just as a transport aid between jobsites.

This R. G. LeTourneau 9-cu-yd type "B" Carryall is matched to a rare Caterpillar Diesel Seventy tractor working on a State Route 76 road project in California, April 1933.

The 8-cu-yd capacity type "H" Carryall introduced in early 1934 was a scraper model that utilized some features of the "B" Carryall, as well as some that would eventually be seen in the new "J" Carryall to be introduced in late 1934. Pictured in early 1934 is an "H" Carryall equipped with Goodyear Airwheels.

R. G. LeTourneau first introduced his four-drum PCU cable control unit in 1933, which allowed a contractor to run a tandem set of Carryalls for increased production. Shown in 1935 is a tandem-set of 12-cu-yd scrapers (type B front, J-12 rear) Carryalls working on the Jack Rabbit Trail, one of the largest earth-moving jobs in Southern California at the time.

Following in the footsteps of the "J-12" Carryall were the smaller "J-6" and "J-8" model scrapers, both in-troduced in early 1935. Pictured in September 1936 is an 8-cu-yd capacity type J-8 Carryall.

Not all of the company's scraper offerings in the 1930s were of a complicated nature. Some were pretty simple in design, such as this little 4.4-cu-yd capacity "Drag-Scraper" from 1935, operated by a single-drum (or larger) PCU, and was built for use on short haul jobs.

After R. G. LeTourneau was able to negotiate the purchase of his original patents concerning the designs of the telescoping scraper from the Warren Brothers Company in mid-1935, he went right to work building a new experimental scraper. Referred to as the type "J-24" Carryall (also known as the 24-Yard Earthmover), it featured a 5-bucket telescoping design. Pictured in November 1935 is the first J-24 Carryall testing at the Peoria, Illinois, manufacturing plant.

The first type "J-24" Carryall would ship in late February 1936, to a jobsite near Ocala, Florida, for work on the Trans-Florida Canal project. Pictured is the 24-cu-yd capacity J-24 at work in Florida in March 1936.

Along with the Peoria-built "J-24" Carryalls, the LeTourneau Stockton, California, facility also assembled one unit in the spring of 1936 for use on the northern approach to the Golden Gate Bridge. It is shown here at work on the U.S. 101 approach on July 1, 1936. *Image courtesy H.C.E.A.*

Shown at work on a Morrison-Knudson railroad project located in Missoula, Montana, on June 18, 1936, is the third J-24 Carryall built. Only three J-24 Carryalls shipped into service. Old company records seem to indicate that a fourth unit was in the Peoria factory inventory as of July 1936, but never shipped. The J-24 was available from LeTourneau on a rental basis only. *Image courtesy H.C.E.A.*

The first "production" Carryall scraper to be produced that benefitted directly from experience gained from the "experimental" J-24 program was the Model "U-18." The U-18 Carryall was the company's first telescopic-bucket scraper design to go into full production after the repurchase of his original patents from Warren Brothers by LeTourneau. Pictured in September 1936 is the Stockton-built pilot U-18 Carryall.

The Model U-18 Carryall was a double-bucket telescoping design rated at 17.7-struck/22-heaped cubic-yards. Shown is the pilot U-18 Carryall at work in October 1936.

The Model "U-18" Carryall set the groundwork for all future multi-bucket scraper production for years to come for the company. It would be replaced by the Model U-20 Carryall in mid-1937. Total U-18 production was listed at 28. This U-18 is working on a dam project near Chattanooga, Tennessee, in February 1937.

The Model "U-20" Carryall was reclassified as a Model "AU" in July 1938. The AU Carryall carried a pay-load capacity of 17.1-struck/24-heaped cubic-yards. Pictured in February 1940 on a roadwork job in Long Island, New York, is a fully loaded Model AU Carryall.

The type "YR-12" Carryall was introduced in mid-1937 and featured a "spring-tube" over the bowl for controlling the tailgate. This YR-12 pictured in May 1939 is on a road job in Vineland, New Jersey.

Capacity for the type "YR-12" Carryall was 11-struck/14-heaped cubic-yards. In July 1938 the YR-12 was reclassified as the Model "K" Carryall. Pictured in May 1939 is an older YR-12 on a roadwork job in Powder Springs, Georgia.

The Model "LU" Carryall introduced in mid-1940 was an expanding double-bucket design rated at 15-struck/19-heaped cubic-yards. Production on this model would end in 1943. Pictured on June 15, 1940, is the first production LU Carryall.

The Model "LS" Carryall introduced in early 1940 was a single-bucket design rated at 8.2-struck/11-heaped cubic-yards. The LS had the longest production run of any of the company's Carryall scrapers.

The LeTourneau Model "DLS" Carryall was a modified version of the company's "LS" Carryall built to fit a Caterpillar DW10 rubber-tired tractor. The first DLS Carryall was produced in June 1941. This DLS and Caterpillar DW10 from November 1941 are working at an airport job in Meridian, Mississippi.

One of the larger capacity Carryalls produced by the company was the Model "NU." Introduced in November 1939, the large double-bucket Carryall was rated at 31.4-struck/42-heaped cubic-yards. Ultimately only two were ever built. Pictured working around the company's Toccoa, Georgia, plant in March 1940 is the huge NU Carryall.

The Model "NP" Carryall was originally introduced in 1942 and was in production for that year only. Payload capacity of the NP was 24-struck/30-heaped cubic-yards. Shown in July 1942 is an NP Carryall leveling an area for the building of LeTourneau steel houses in Tournapull, Georgia.

Introduced in February 1939, the Model "SU" Carryall featured a double-bucket design capable of handling a 14.3-struck/18-heaped cubic-yard payload. But wartime restrictions would cause the model line to be cancelled in 1941. This SU Carryall is pictured in October 1939 on a road working job in Belleville, Illinois.

LeTourneau introduced its very popular Model "W" Carryall in December 1938. The single-bucket W was rated at 18.7-struck/23-heaped cubic-yards. In August 1939 the model would get an updated "streamliner" gooseneck yoke arrangement. Production would end on the W Carryall in 1952. Pictured is a W Carryall working on a dam project in Des Moines, Iowa, on July 14, 1944.

The Model "FU" Carryall introduced in the fall of 1940 featured a roller-mounted, double-bucket design which allowed the bowl to expand with less rolling resistance as in previous "U" series scrapers. Capacity was listed at 17.7-struck/23-heaped cubic-yards. Shown is a Model FU on a jobsite in El Toro, California, in December 1940.

The original Model "RU" Carryall was introduced by LeTourneau in September 1938. It would receive the new "streamliner" gooseneck yoke upgrade in June 1939. Pictured in January 1942 is one of the updated RU Carryalls on an airport job located in Goleta, California.

The towed Model "RU" Carryall was a double-bucket design rated at 23-struck/30-heaped cubic-yards. Only the most powerful track-type tractors of the day were capable of handling the RU (or the company's own Tournapull rubber-tired tractor unit). Image date June 1945.

In the early 1950s LeTourneau started to introduce all-new Carryall models featuring an open-bowl design. Models of a "PCU" cable-controlled nature were the O-14, O-19, O-23, and O-35, all in 1952. Models that utilized scraper-mounted electric motors for the unit's cable-control functions were the PO-14 (early 1953), and the PO-19 and PO-23 (late 1952). Shown at work in April 1953 is a Model PO-23 Carryall near the LeTourneau Vicksburg, Mississippi, assembly plant. The PO-23 was rated at 14.4-struck/19-heaped cubic-yards in capacity.

The LeTourneau Model "A" Tournapull and "Z25" Carryall was the earth-moving industry's first high-speed, rubber-tired, self-propelled scraper. Completed at the Peoria, Illinois, facility in April 1938, it is shown here at the company's test farm not far from the main plant in May 1938.

The history making "A" Tournapull was powered by a 160-hp, 8-cylinder Caterpillar D17000 diesel engine. Its experimental "Z25" Carryall was rated at 25 cu-yds in capacity. Image taken at the company's test farm in May 1938.

The prototype Model "A" Tournapull and "Z25" Carryall would not be placed into service, at least not as a complete unit. The front Tournapull tractor became part of the LeTourneau Cane Harvester that would ship from the factory in February 1939 to the Hawaii Sugar Planters Association for evaluation. Pictured is the completed harvester in Peoria in December 1938.

The next Tournapull built by the company was the series "A-A" featuring a revised chassis design for the tractor unit and a new "TU" Carryall rated at 17.1-struck/24-heaped cubic-yards in capacity. The first Model A (A-A) Tournapull is pictured here in July 1938 testing at the Peoria facilities.

Shown in August 1938 is a Model "A" (A-A) Tournapull equipped with a "TU" Carryall at the company's test farm. Total production of Model A (A-A) Tournapulls was only five units.

After a successful test of a Model "A" (A-A) Tournapull by Guy F. Atkinson Co. at their Hanson Dam project near San Fernando, California, the contractor placed an order for ten Model "A1" Tournapulls equipped with an improved "HU" carry scraper. The HU Carryall was a double-bucket design rated at 22.2-struck/30-heaped cubic-yards in capacity. These new models would start to work in October 1938. Pictured at the Hanson Dam project in December 1938 is one of the new A1 Tournapulls.

Early on the fleet of Tournapulls working at the Hanson Dam project was experiencing premature tire and wheel-rim failures. Pictured in August 1939 is one of the Guy F. Atkinson A2/RU Tournapulls equipped with redesigned wheels. LeTourneau first started shipping the improved A2 Tournapull with the RU Carryall in December 1938. *Image courtesy Ed Brooking/R. G. LeTourneau Hertage Center.*

Guy F. Atkinson Co. operated the largest fleet of the early "A" Tournapulls, 14 in all, all at the Hanson Dam jobsite. They included one series A-A, ten series A1, and three A2 Tournapulls. Pictured in August 1939 is another of Atkinson's A2/RU models equipped with revised wheels and tires. *Image courtesy Ed Brooking/R. G. LeTourneau Hertage Center.*

The tire wear problems on the early Tournapulls led the company to experiment with larger tire and wheel combinations. Pictured is a Model "A3" Tournapull (with RU Carryall) equipped with experimental Firestone-built, "balloon" tires on the front, mounted on reinforced rims. Image taken at the Santee Cooper dam project in South Carolina in December 1939.

The first "production" Model "A3" Tournapulls started rolling off the company's Toccoa, Georgia, assembly line in September 1939, equipped with RU Carryall scrapers. The A3 featured a big Caterpillar D17000 diesel engine, and was fitted with larger 30x40-ich tires designed by LeTourneau and manufactured by the Firestone Tire & Rubber Company. Pictured in February 1940 is a Peoria-built Model A3 Tournapull mated to a huge "NU" Carryall rated at 31.4-struck/42-heaped cubic-yards.

Unfortunately, the Model "A3" Tournapull never performed up to expectations for the company. All together 20 Model A3 Tournapulls were built, with 11 eventually being returned to the Toccoa, Georgia, plant and disassembled. Pictured is an A3 Tournapull in September 1940, featuring an updated RU Carryall design.

The first of the twin-engine powered Tournapulls to be introduced by LeTourneau was the Model "A5." Introduced in February 1941, the A5 Tournapull was equipped with two six-cylinder, Cummins HBIS600 super-charged diesel engines rated at 400-hp combined. Each engine utilized its own transmission and final drive assemblies connected by hydraulic couplings. Pictured in July 1942 operating near the Toccoa plant is the A5 with an OU Carryall.

The Model "A5" Tournapull was originally introduced equipped with an "NU" Carryall. In January 1942 the model was fitted with a huge "OU" Carryall rated at 41-struck/60-heaped cubic-yards in capacity. Only one A5 Tournapull was ever built, and it is shown here in July 1942 working not far from the main Toccoa, Georgia, assembly plant.

Following the "A5" was the twin-engine Model "A6" Tournapull in March 1942. Built at the Toccoa plant, the A6 was equipped with the Model "OU" Carryall first seen with the A5 Tournapull. Tires on the front tractor were large 36x40-inch units co-developed by LeTourneau and Firestone (the same as the A5). Pictured operating near the Toccoa factory is the pilot A6 Tournapull. Image date late March 1942.

The Model "A6" was powered by the same model type Cummins diesel engines found in the earlier Model "A5" Tournapull. The transmission however was a LeTourneau-designed power-shift, four-speed "Tournamatic" with air-assisted push-button and lever controls. Image taken in late March 1942.

What do you get when you take one "A6" Tournapull and attach it to a 75-ton capacity, heavy-duty trailer? Why a Model "A" Tournatruck of course! LeTourneau built a total of four Model A Tournatrucks, with the first prototype pictured here in April 1942 hauling a test load at the Toccoa, Georgia, assembly plant.

Shown at the Toccoa, Georgia, plant in September 1942 is the LeTourneau Land Battleship concept. Comprised of two "A6" Tournapulls, the Land Battleship would have carried up to a 155 mm howitzer in size over rough terrain in a combat situation. During demonstrations the concept was fitted with a mock-up of a cannon to get the message across to Army brass and on-lookers alike. It would never go into production. Total number of all A6 tractors was only nine units (including the two utilized in the Land Battleship).

Last of the early twin-engine powered Tournapull designs was the experimental Model "A7." Modern in looks for its time, it is shown here in December 1942, at the Peoria plant, displayed next to the pilot Model "D" Tournapull.

The Model "A7" Tournapull was tested around the Peoria, Illinois, assembly plant in January 1943 (pictured), and was equipped with a Model "NU" Carryall. Drivetrain package was the same as that installed in the previous Model "A6" series. But war-time production needs at the plant would cause the A7 Tournapull project to be shut-down. Only one A7 was ever built.

LeTourneau built a number of experimental "A" Tournapulls in the late 1940s featuring electric-motor cable control for the Carryall scrapers. One of those models was the "A20" Tournapull. Introduced in early 1948, the Model A20 utilized the company's new "Tournapower" drivetrain consisting of a 450-hp Allison V1710 engine (configured to run on liquid butane) and a four-speed "Tournamatic" transmission. Shown at the Vicksburg, Mississippi, plant in October 1950, is an A20 Tournapull with an experimental "E-60" Carryall (originally built in March 1948) featuring a revised gooseneck design. Only five Model A20 Tournapull tractors were ever built.

The Model "A23" Tournapull (with the E-50 Carryall) was originally introduced in November 1950, and was equipped with the 450-hp "Tournapower" drivetrain. In July 1951 the designation of the series was changed to the Model "A" Tournamatic Tournapull. With that came a new "P-35" Carryall scraper choice. Total number of A23 Tournapull tractors built was 11. For the A Tournamatic it was 48. Pictured at the Vicksburg plant in August 1952 is a Model A Tournamatic Tournapull and P-35 Carryall.

Many of the company's Tournapull tractors were offered with heavy-duty "Tournarocker" trailers as an option. Testing with Model "A" Tournapulls started with the company in December 1948 (with the Model A20). Shown at the Vicksburg plant in February 1950 is an A20 Tournapull equipped with a low-sided 35-ton capacity Model "A" Tournarocker (later reclassified as an E-35).

The Tournapowered Model "A20" Tournapull with its Tournarocker was a rather large rear-dump hauler for its day. Pictured is the "A" Tournarocker in late February 1950 during a demonstration at the Vicksburg facilities. Mr. R. G. LeTourneau is the one standing next to the tire in the dark suit.

Shown in April 1951 at the Vicksburg plant is a Model "A23" Tournapull equipped with a Model "E-50" Tournarocker. Note the higher sides of the E-50 required for the trailers 50-ton payload capacity.

This Model "A" Tournamatic Tournapull with its 50-ton capacity "E-50" Tournarocker is working at the Bagdad Copper Mine in Arizona, in May 1953.

LeTourneau built its first Model "B" Tournapull in late October 1938. The original B Tournapull featured a fully enclosed, automotive styled operator's cab, and was equipped with a Model "BU" Carryall rated at 9.3-struck/12.5-heaped cubic-yards. Pictured in Peoria in early November 1938 is the only B Tournapull of this type built.

Another one-off design for the company was the Model "B" (B-1) Tournapull. Completed at the Peoria factory in March 1939, the B (B-1) featured a very streamlined automotive styled tractor design. First scraper of choice was the Model "SU" Carryall rated at 14.3-struck/18-heaped cubic-yards. By May 1939 a different Model "P" Carryall had been fitted to the unusual looking Tournapull (pictured).

In 1939 LeTourneau introduced another "B" Tournapull model in the form of the "B1." The Model B1 Tournapull (not the same as the previous B-1) was designed around a 157-hp, six-cylinder GM diesel engine powered drivetrain. Carryall scraper of choice was the Model "SU." The first B1 Tournapull shipped from Peoria in June 1939. Total number of B1 Tournapulls built was only ten. Image date December 1939.

This image from June 1940 shows a Cummins diesel powered Model "B5" Tournapull on the left, and a Caterpillar D13000 diesel engine equipped Model "B4" on the right. The B5 was fitted with an "LU" Carryall, while the B4 made do with an "SU" unit. Both of these experimental prototypes were designed and built at the Toccoa, Georgia, factory. Only one of each of these Tournapulls was ever built.

The Model "B6" Tournapull was another of the company's "B" tractor designs to be powered by a 140-hp Caterpillar D13000 diesel engine. Scraper choice for the B6 was a Model "LU" Carryall. The first B6 was completed at the Toccoa plant in June 1940. Only three B6 Tournapulls were produced. Image date August 1940.

Of the early mechanical cable-controlled "B" Tournapulls, only the Model "B8" could be considered a "production" machine. Originally completed in November 1940, the Model B8 Tournapull (also referred to as a Super-B) was powered by a 200-hp Cummins HBIS600 diesel engine. Scraper of choice was the double-bucket "LU" Carryall. Total number of standard B8 tractors built was listed at 16 units. There were also four additional Model "B8-A" Tournapulls produced for military purposes. Image date February 1941.

LeTourneau produced a number of experimental and limited production "B" Tournapulls featuring electric control systems in the late 1940s and early 1950s. One of those was the Model "B" Roadster. The B Roadster Tournapull was equipped with a standard 240-hp, eight-cylinder Buda 8-DC-1125 diesel engine mated to a five-speed, heavy-duty, sliding gear transmission. The first B Roadster shipped from the Vicksburg, Mississippi, plant in August 1948. Total number of tractors built was listed at 42. Pictured at Vicksburg in December 1950 is a B Roadster equipped with a 25-ton capacity E-25 Carryall.

The most popular Tournapull series ever produced by the company was without a doubt the Model "C." Originally unveiled in April 1940, the pilot C Tournapull is pictured here at the Peoria plant in May 1940 undergoing preliminary field testing trials.

LeTourneau made a few changes to the original design of the Model "C" Tournapull including a modification of the tractor chassis. The C Tournapull was powered by a 90-hp, 1,800-rpm, six-cylinder Caterpillar D468 diesel automotive engine mated to a four-speed Fuller gearbox. Shown in July 1940 is one of the first C Tournapulls (suffix code C1) built featuring the new changes to the line.

Carryall of choice for the early Model "C" Tournapulls was the "LS" scraper rated at 8.2-struck/11-heaped cubic-yards in capacity. This C Tournapull from November 1940 is working on a road rebuilding project on a Washington state highway located near the Grand Coulee Dam.

For contractors needing a more powerful version of the popular scraper, LeTourneau offered the Model Super "C" Tournapull. Standard engine was the 150-hp Cummins HBID600 diesel, with optional 150-hp Buda 6DH691, and 124-hp Hercules DRXC powerplants also available. The first pilot Super C Tournapull was completed at the Peoria factory in late October 1940. Pictured is a Super C working on the Albany Army Airfield construction project in Georgia, in the spring of 1941.

The Model Super "C" Tournapull was equipped with an "LP" Carryall scraper unit rated at 12.1-struck/15-heaped cubic-yards. These Super C Tournapulls pictured in April 1941 are operating in Albany, Georgia, on an Army airfield project.

Super "C" Tournapulls could really move the dirt. With their basic diesel engines and manual four-speed Fuller 4-B1-86 model heavy-duty gearboxes, they were as reliable as a self-propelled scraper got in the 1940s. Here a Super C is loading with the assist of a pusher dozer at Camp Pendleton, California, on November 25, 1945.

Problem areas associated with the Super "C" Tournapull were few. One area of concern was the catastrophic failure of the "C" hitch at speed which would most certainly ruin the operators day. Another area was the steering clutches which could get a bit "squirrely" if applied too abruptly during adverse working conditions. Image taken in March 1946.

The Model Super "C" Tour-
napull was one of the most
recognized self-propelled
scrapers of its day, and is
considered a true milestone
in the history of earth-moving equipment overall. It is
a "classic" in every sense of
the word. Image date May
1946.

The Model "C" and Super "C" Tournapull tractors made ideal platforms for uses other than as a scraper.
Shown in March 1944 is a Super C Tournapull equipped with a LeTourneau "W210" Tournatrailer. Originally
introduced in 1942, the W210 was a rear slide-out dumping design rated at 11.5-struck/17-heaped cubic-
yards in capacity.

Another variation on the Model "C" Tournapull theme was the Tournalift. Pictured working at an Army Depot in 1943 is a Super C Tournapull repowered with a Caterpillar diesel engine, equipped with the rear-facing fork-lift unit.

This version of the Tournalift features a downsized Model "C" Tournapull and a smaller rear-mounted fork-lift attachment.

This one-of-a-kind LeTourneau creation was simply referred to as the "Experimental Bulldozer and Trailing Wheel." Utilizing a Super "C" Tournapull, the tractor seems to have been designed to perform work around the Peoria manufacturing plant. Image date February 1944.

LeTourneau first developed the concept of a two-powertrain, high-powered tow-tractor unit in late 1951 in the form of the "Tournatwo." The company produced two variations on the theme—the larger Model "A" and the smaller Model "C" Tournatwo. The first to start development, the C Tournatwo was powered by two 186-hp GM 6-71 diesel engines (372-hp combined). Transmissions of choice were Tournamatics. Four complete C Tournatwo tractors were eventually built at the Longview, Texas, plant in 1952.

In 1942 the Engineer Board of the U.S. War Department made a request to LeTourneau to design an air-transportable Carryall scraper and retrofit hardware to be installed on a Willys Jeep. The company's answer was the Model "C" Jeep Scraper. But the special one-off C Carryall was too much for the Jeep to ultimately handle and a better solution would need to be found. Shown is the Jeep Scraper at the company's Peoria test farm in early September 1942.

The first Model "D" Tournapull was completed by LeTourneau at the Peoria plant, in December 1942 (pictured). The pilot D Tournapull was originally equipped with a Model "Q" Carryall rated at 2 cubic-yards heaped in capacity. A week or so after this image was taken the prototype D'pull had its Carryall scraper replaced with a flatbed "Tournatruck" trailer.

Shown at the Peoria plant in late December 1942 are the first two Model "D" Tournapulls assembled. On the left is the prototype D Tournapull (S/N C5T-2001-D) now equipped with a "Tournatruck" trailer, and on the right is the second unit (S/N C5T-2002-D). The number two D'pull is fitted with the original "Q" Carryall scraper unit that was first tested with the pilot tractor earlier in the month.

The Model "D" Tournapull (sometimes referred to as the "Flying Tournapull") was powered by a 44-hp, four-cylinder Continental Y112 gasoline engine. Transmission choice was a manual four-speed Borg-Warner gearbox. Image taken in Peoria, March 1943.

This rather used looking Model "D" Tournapull and Tournatruck is taking on fuel at the main plant in Peoria, and was used primarily to shuttle supplies around the factory grounds. The unit is most likely the pilot D'pull since records indicate that it never left the Peoria facilities. Image date November 1943.

In late December 1943 the company completed a one-off "D3" Tournapull equipped with the "Q" Carryall (pictured). This experimental unit featured an elevated operator's seat and controls for better visibility. Image taken in the first week of January 1944.

The small "airborne" scraper model to actually get the go ahead to proceed into production was the "D4" Tournapull. Equipped with the "Q" Carryall, the Model D4 looked very much like the earlier D test units, but featured many refinements to the overall design. The first D4 was completed in late December 1943, and was released for shipping from the Peoria plant on January 8, 1944. Pictured at the Toccoa plant in September 1944 is a Model D4 equipped with the standard front-mounted "AD" Tiltdozer.

Improvements were also made to the "D4" Tournapulls "Q" Carryall design as well. The little Carryall now had extended sides which raised the capacity slightly to 2-struck/2.3-heaped cubic-yards. Other upgrades included additional tailgate return springs and stronger rear wheels. Image taken at Peoria, March 1945.

In December 1944, LeTourneau officially announced the availability of the Model "D4" Tournapull to the civilian marketplace. These two D4 Tournapulls are shown working on a road job in Greene County, Georgia, in April 1945.

The Model "D" Tournapull was produced in numbers that were considered reasonable for such a specialized machine. Total number of the early "D" models built was 10 units, with the more popular D4 version coming in with 636 (including model variations). Last year of production was in 1946. Pictured in August 1945 is a D4 Tournapull on a roadwork job in Gainesville, Florida.

The snub-nosed Model "D6" Tournapull featured electric operator controls and motors for the cable controls on the scraper unit. Power for the D6 was supplied by an 85-hp Buda HP-351 gasoline engine. Standard Carryall choice was the Model "E-4" rated at 3.3-struck/3.7-heaped cubic-yards. First "production" D6 shipped in February 1947. Total number built was 93 units. Picture taken in Toccoa, March 1947.

Mr. R. G. LeTourneau poses next to a Model Super "D6" Tournapull in this publicity image taken in Peoria, Illinois, April 1947. The Super D6 was powered by a 113-hp Buda 6-DT-468 diesel engine.

Shown at the Peoria plant in January 1950 is a Model "D" Roadster Tournapull equipped with an "E-9" Carryall rated at 5.9-struck/7-heaped cubic-yards. Engine in this model Roadster was a 122-hp GM 4-71 diesel. The D Roadster Tournapull originally started rolling off the Peoria assembly line in the summer of 1948.

R. G. LeTourneau first fitted a bull-dozer like blade attachment to a Best "Sixty" tractor in the spring of 1926, while working on an earth-moving contract at Crow Canyon in California. His next try would be a more refined design utilizing his new cable-control system on a job contract in Oroville, California, in the summer of 1928. He would officially start to offer his new "Bull-Dozer" tractor attachment to other contractors starting in 1929. Pictured in 1932 is a Caterpillar "Sixty" equipped with one of LeTourneau's early production bulldozer blades.

Shown at the company's Stockton, California, facilities in 1932, is a Caterpillar tractor equipped with a LeTourneau Bulldozer and rear-mounted Power Control Unit (PCU). LeTourneau also offered bulldozers for the Allis-Chalmers Model "L" and the Cletrac "80" track-type tractors.

In early 1932 LeTourneau introduced a rear-mounted tractor attachment for the Caterpillar "Sixty" referred to as the Tracked-Mounted Scraper, more commonly known as the "Cowdozer." The Cowdozer was best utilized in ditch digging and hill side grading applications. Only five are listed as being produced. Image date 1932.

This rare Caterpillar Diesel Seventy, equipped with a LeTourneau Angledozer and rear-mounted PCU, is working on a section of U.S. 101 in Las Flores Canyon, California, in April 1933.

Shown in the summer of 1933 is a Caterpillar Diesel Fifty equipped with one of the first "production" LeTourneau Angledozer blade attachments.

The early LeTourneau bull-dozer blades were designed to take on a variety of tough dozing applications, such as this Angledozer unit attached to a Caterpillar Diesel Fifty working on a Sequoia National Park road job in September 1933.

Shown in August 1937 at the Peoria plant is a Caterpillar RD6 equipped with a rather complicated looking Model AFC6T Angledozer (sometimes referred to as a "Streamlined" Angledozer). It also features a front-mounted PCU blade control.

The LeTourneau Model "E" Bulldozer was produced in models to fit the Caterpillar D4, D6, and D7 track-type tractors. Pictured is an "E4" Bulldozer fitted to a D4 working on a section of the U.S. 85 highway project near Denver, Colorado, in October 1939.

Equipped with a Model "CK8" Angledozer, this Caterpillar D8 is at work on part of the U.S. 27 road job near Rome, Georgia, in April 1940.

Pictured in the Peoria factory yard in December 1942 is a Caterpillar D8 equipped with a Model CK8 "Engineer-Standard" Angledozer, soon to see military action overseas. LeTourneau produced their 5000th Angeldozer in February 1942.

LeTourneau also produced "pusher" attachments for the larger track-type tractors to aid in scraper loading. The Model "CP" Pushdozer was available as a complete package, or as a "Pushdozer Cup Group" option that could be installed to any Model CK8 or C Angledozer to convert them into a pusher. Pictured in July 1940 is a Model CP7 Pushdozer mounted on a Caterpillar D7.

LeTourneau also designed bulldozing blades for specialized purposes, such as this rather wicked looking "Rootdozer" working on a land clearing job near the Toccoa, Georgia, manufacturing plant, in February 1940.

Shown is one of the earliest examples of an R. G. LeTourneau designed crane fitted to the rear of a Caterpillar Sixty tractor. Note the sandbags at the front of the tractor for ballast. LeTourneau operated this unit while working on part of Southern Pacific's Suisun Bay railroad bridge project in Benicia, California, in the fall of 1929.

This wide-gauge Caterpillar D6 shown at LeTourneau's Peoria factory yard is equipped with a Model "WE6" Bulldozer, and a rear-mounted Hyster crane/winch attachment.

The first rubber-tired dozer prototype produced by LeTourneau was the Model "T-200" Tournadozer (S/N SP-T-2095-T200, pictured). First unveiled at the Peoria factory in November 1945, it was powered by a 300-hp Buda 8-DCS-1125 super-charged diesel engine.

The pilot "T-200" Tourna-dozer shipped from Peoria in December 1945 to Vicksburg, Mississippi, to aid in an airport job close to the company's factory there. After that, it was shipped to the Toccoa, Georgia, factory to work on another airport project in the summer of 1946. Pictured is the T-200 Tournadozer (with modifications) at work in Toccoa.

Largest of all the rubber-tired dozers designed by the company was the Model "A" Tournadozer. Built at the Longview, Texas, plant, it would start field trial testing in September 1947. Pictured in June 1948 is a Model A Tournadozer at a levee earth-moving jobsite near Prairie Du Rocher, Illinois.

The Model "A" Tournadozer was originally powered by a 750-hp Packard 4M2500 marine diesel engine configured to run on liquid butane. But the tractor's power output was too much for its 4-speed Tournamatic transmission to handle, so the engine was de-rated to 500-hp. Shown in August 1948 is the Prairie Du Rocher, Illinois, machine push-loading a Tournapull equipped with an "E-35" Carryall scraper unit.

The 35,000-pound Caterpillar D8 track-type tractor was considered large in its day, but it is dwarfed in size when parked next to a 93,000-pound Model "A" Tournadozer. But the reliability issues with the "A" dozer could never be overcome by the LeTourneau engineers, and only a handful were ever produced. Image date August 1948.

The second prototype Tournadozer produced in Peoria was completed in August 1946 and started testing the following month. This tractor originally started out life being referred to as a Model "T-200," but was later reclassified as a Model "B" Tournadozer. Pictured is this unit testing in Peoria in June 1947, now equipped with a redesigned operator's station.

The Model "B" Tournadozer would go into limited production starting in mid-1947. Engine of choice was a 300-hp Super Buda Diesel Model 8-DCS-1125. The B Tournadozer weighed approximately 50,700 pounds, compared to the smaller C Tournadozer's 31,000-pound shipping weight. Pictured in September 1947 is a Longview-built B Tournadozer.

The first rubber-tired dozer to be built at the company's Longview assembly plant in Texas was the pilot Model "C1" Tournadozer. Shown in Longview in November 1946 is the prototype C1 Tournadozer. By the end of December redesigned air-actuated disc clutch and brake system controls would be installed.

Shown on the Longview assembly line in late January 1947 is the first "production" Model "C1" Tournadozer nearing completion. On February 5, 1947, this C1 was released for shipping up to the Peoria plant for further testing and engineering evaluations.

The first production Model "C1" Tournadozer was powered by a 160-hp Buda 6-DC844 diesel engine mated to a 4-speed Tournamatic transmission. Pictured in Peoria in February 1947 is the first production C1.

In mid-1947 LeTourneau introduced an upgraded "C" Tournadozer in the form of the Model "C2." The new C2 Tournadozer featured a redesigned front A-frame, and a new standard wheel and tire combination. Power had also increased to 180-hp. Shown in May 1947 is the first C2 Tournadozer off the Longview assembly line.

The most popular rubber-tired dozer produced by LeTourneau was by far the Model Super "C" Tournadozer. Introduced in late 1948, the Super C was equipped with a 210-hp Buda 6-DA-844 diesel as the standard engine choice. Pictured at the Longview plant in December 1948 is the first Super C Tournadozer ready for shipping.

In mid-1952 LeTourneau started offering a special model of the Super "C" identified as the "Tournatractor." The Super C Tournatractor was equipped with a heavy-duty front bumper that also doubled as a counterweight, and a rear-mounted PCU for scraper towing applications. Shown in Longview in mid-1952 is an early version of the Super C Tournatractor.

Introduced in June 1952, the "Tournadiggster" was a front-end loading shovel featuring electric motor driven hydraulic pumps for hoist and crowd functions. The attachment (built for the Super C Tournatractor) was developed jointly between R.G. LeTourneau's Machinery Division and Dempster Brothers, Inc. of Knoxville, Tennessee. Image date July 1952.

R. G. LeTourneau first introduced his 16-cu-yd capacity "Dump-Cart" wagons in the spring of 1928 (Fresno, California, contract). These early hopper wagons rode on large welded steel wheels, but in late 1928 a version featuring a track-design ideal for soft working conditions was introduced (Newhall, California, contract). Pictured in early 1929 is one of these Dump-Carts featuring the first track configuration designed by LeTourneau. A more refined track design was introduced in the summer of 1929 (first used at the Benicia, California, contract). *Image courtesy of KAC.*

LeTourneau originally introduced its slide-out dumping, rubber-tired hauling trailer in late 1933 as the "25-Yard Buggy." The LeTourneau Buggy was eventually available in two variations; 18-struck/24-heaped-and 23-struck/30-heaped cubic-yard models. Depending on the working conditions the early buggies could be specified with eight high-floatation "airwheels," or 16 tires for harder packed working areas. Pictured is an eight-wheel, 30-heaped cu-yd Buggy hauling material on the Bonneville Dam spillway project on the Columbia River, in Washington, in early 1935.

Shown is a Caterpillar Diesel Seventy-Five track-type tractor with a standard model 24-heaped cubic-yard, 16-wheel LeTourneau Buggy in tow modified with side-boards added to increase capacity, working on a canal project in Los Angeles, California, in 1934.

In early 1935 LeTourneau produced a special order truck and buggy hauling unit for Kaiser Construction for use on the Bonneville Dam project on the Columbia River. The tractor utilized on the rig was a modified Mack AP unit. Pictured at the Stockton, California, factory testing area is the LeTourneau Truck Buggy. *Image courtesy H.C.E.A.*

The buggy utilized on the tractor/trailer rig was a modified version of a track-type tractor towed model from 1934. The 16-wheel buggy utilized with the Mack AP truck was rated at 35-struck cu-yds, with a 61-ton payload capacity, and was a slide-out dumping design. Photo taken early 1935. *Image courtesy H.C.E.A.*

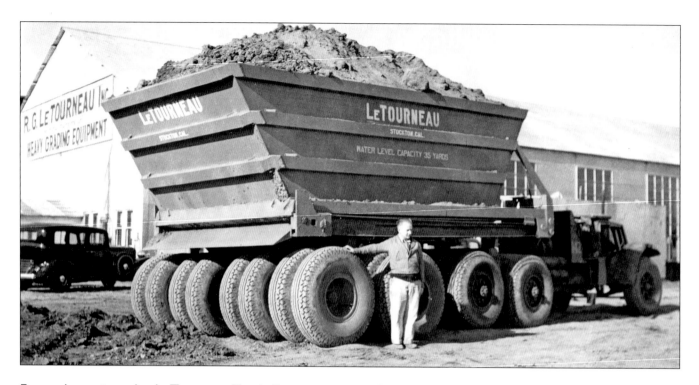

For a short time the LeTourneau Truck Buggy was considered the world's largest dump truck unit when it went into service for Kaiser Construction. With 26 tires on the ground it was a real news maker in the trade magazines in 1935. Photo taken early 1935. *Image courtesy H.C.E.A.*

The LeTourneau "Cradledump" was first introduced in late 1935. Designed as a cable-controlled side-dump, it was rated at 30-struck/35-heaped cubic-yards in capacity. It was offered equipped with a dolly for track-type tractor use, or with a fifth-wheel hitch, like the one shown attached to a Caterpillar diesel powered Hug Lugger Model 100 tractor unit. Picture taken in Peoria in December 1935.

In 1936 The Hug Company (of Highland, Illinois) started to offer a couple of Caterpillar diesel engine powered truck chassis' equipped with LeTourneau "Slide-Out Body" units. Pictured in December 1936 is a Hug Lugger Model 30 fitted with a 10-cu-yd LeTourneau Slide-Out Body and front-mounted PCU. *Image courtesy H.C.E.A.*

The LeTourneau 15-Yard Buggy was originally introduced by the company in the summer of 1937. It would be reclassified as a Model "R-15" in October 1937. Pictured in June 1937 is the first design of the slide-out dumping 15-Yard Buggy with the original wheel configuration, at the company's Peoria test farm.

The Model "R-15" trailer would have its wheels re-mounted to each side of the main axle-supports for better stability. Shown is an updated R15 unit working on a section of the U.S. 40 project near Frederick, Maryland, in November 1937. In March 1938, the R-15 was reclassified as the Model "R-18."

The earliest tractor-towed ripper invented by R. G. LeTourneau was his "subsoiler" (also referred to as a scarifier or rooter) from 1920, and was built strictly for agricultural use. The subsoiler would eventually pass into the hands of Ephraim (Eph) Hahn, who had operated the "Mountain Mover" for LeTourneau previously from 1922 to 1926 (who would also become the owner of that piece of equipment as well). Pictured is the subsoiler with Eph's wife sitting in the operator's seat of the Holt tractor. *Image courtesy R. G. LeTourneau Heritage Center.*

The next ripper to be built by LeTourneau was a dual rack-and-pinion, electric-motor controlled subsoiler (utilized with his first bull-dozer equipped tractor) operated at his Crow Canyon contract in the spring of 1926. This was followed by the more simplified design seen here in the summer of 1926 that was utilized on the Philbrook Dam project located in Butte County, California. *Image courtesy KAC.*

R. G. LeTourneau first introduced his cable-controlled, multi-shank "Hard-Pan Rooter" in 1929. LeTourneau had previously built a single-shank prototype similar to his cable-controlled design that was pneumatically operated (spring 1928, Southern Pacific railroad freight yard contract, Fresno, California). Pictured is a circa 1930 LeTourneau heavy-duty ripper, now referred to simply as a "Rooter."

This unique looking piece of equipment is a PCU operated, eight-wheel LeTourneau Logging Arch. Introduced in 1935, it was built for The Red River Lumber Company, and was the first known use of dual tires on a logging arch design. *Image courtesy H.C.E.A.*

Chapter 2 – The Electric-Drive Wheel Becomes a Reality

The first few decades of production for R.G. LeTourneau, Inc. saw heavy machinery move from steel wheels to pneumatic tires, and mechanical cable controls replaced by electric motors. But R. G. LeTourneau the inventor still had a few more "tricks" up his sleeve, namely the combining of the wheel and an electric drive motor into a single assembly. Now driving a wheel with a shaft through an axle housing with an electric motor mounted inside was nothing new, but combining the pair was. By mounting the traction motor in the wheel assembly itself, he had created what was essentially an electric drive wheel. R. G. would first show off this new design concept at a special event staged at the Edgewater Beach Hotel in Chicago on November 17, 1950. There, he demonstrated a vehicle referred to as a "Tournatow" equipped with four electric drive wheels. Though the Tournatow had little sales impact, the technology it showcased, namely the electric drive wheel concept, would go on to refocus the company's engineering vision as to designs that would utilize this new form of propulsion.

On May 1, 1953, R. G. LeTourneau finalized the sale of the earth-moving equipment lines of his company, along with key manufacturing facilities that built them, to Westinghouse Air Brake Company, a division of the Westinghouse Corporation of Pittsburgh, Pennsylvania, forming the company LeTourneau-Westinghouse. The sale of these parts of his company left R. G. with sufficient funds to further develop the electric-drive wheel concept to make it more reliable and acceptable in the marketplace. Just about every land-based heavy-equipment design to be put forth by the company from mid-1953 onward would utilize the revolutionary drivetrain powered by diesel, gasoline, or butane fueled engines.

During this time period R.G. LeTourneau, Inc. produced an astonishing amount of equipment designs utilizing the company's electric-drive wheel drivetrain layout. The incredible array of prototype and limited production machinery built was simply astounding. Massive land trains, missile launchers, jungle crushers, multi-bowled scrapers that could carry 360 tons of earth in a single trip, the list goes on and on. Many of the machines looked as if they had driven right out of a science fiction novel. These designs had never been seen before in the marketplace before and potential buyers were hard to come by. The building of so many concept machines today from a single manufacturer would be unthinkable, as well as financially irresponsible. But at the time it was R. G.'s company and he ran it as he saw fit.

Looking back at all of the varied product lines and specialized equipment designs produced in the late 1950s and all of the 1960s by the company, a few would succeed remarkably well and stand the test of time, such as the log stackers and the off-shore jack-up rigs. But for most it was the end of the line with the vast majority of the experimental machines meeting their ends in the furnaces of the company's Longview steel mill in late 1966 and throughout 1967, to be reborn again in future LeTourneau creations.

R. G. LeTourneau points out a few of the technical highlights of his revolutionary electric drive-wheel concept on his latest new model, the "Tournatow," at its official unveiling at the Edgewater Beach Hotel in Chicago, Illinois, on November 17, 1950.

Shown in service at MacDill Air Force Base in Florida is a Model A-3 (military designation Model AT-4) "Air-Tow." The designation of the model line was changed from Tournatow to Air-Tow after the sale of key assets of the company to LeTourneau-Westinghouse in May 1953. Two of these special Air-Tows were built (S/N 3674/3675) and both shipped to Mossy Head, Florida, on September 4, 1953.

Built for the U.S. Navy Bureau of Docks, the "Bu-Dock" (also known as the "Gilhoist Tows Underslung Tractor LC-23") was powered by a Buda 6-DAS-844 supercharged engine. Shown is the prototype unit on December 1, 1954. Only two of these tractors were ever built (S/N 5194/5195).

The Series PT-1 (military designation Model FTR) Fighter Trailer was designed to haul fighter jets utilizing tricycle landing gear layout. The trailer utilized non-powered wheels and was designed to be towed behind another vehicle. The prototype PT-1 (S/N 5266) was shipped to Edwards Air Force Base in California on February 4, 1956, and it is pictured here loaded with an F86F Sabre on March 26, 1956.

Another military transport vehicle designed by R.G. LeTourneau was the Series "PT-3" Jet Fighter Transporter. Eighteen of these transporters were built for the Danish Air Force (RDAF) in 1962. Pictured is one of the three-wheeled PT-3 units transporting an F100D.

Mistakenly described as a military vehicle in one of my previous books, the Model "E18" Tournashuttle is in fact an underground ore shuttle car built for Turner Construction Company, for use at the White Pine Copper Co. mine, located in White Pine, Michigan. Three of these shuttles were built (S/N 3801/3802/3803) with the first unit (pictured) officially released for shipping on May 21, 1953. *Image courtesy H.C.E.A.*

The Series "MT-1" Scout Rocket Transporter was originally built for use by Vought Astronautics in Dallas, Texas, the primary contractor of the rocket for NASA. The first Scout (Solid Controlled Orbital Utility Test system) transporter S/N 6011 (pictured) shipped to Vought in December 1961. A total of six MT-1 trailers were eventually built.

The A-Tow No. 1 "Army Plow" was originally built in 1952. It was shipped to the U.S. Army Engineer Property Officer, Fort Know, Kentucky, on March 15, 1952, to start its initial testing program. The mine plow's (nicknamed Pistol Pete) electric drive system was powered by a 500-hp Allison V-1710-111 engine, and was capable of clearing a path 14 feet wide, 3 inches deep. Pictured in 1953 was the only unit built (S/N 3595).

The Model TC-264 "Sno-Buggy" was originally built to the requirements made by the U.S. Army Transportation Corps for a vehicle capable of traveling across the Greenland Ice Cap towing cargo trailers and sleds. Completed in June 1954, the TC-264 (S/N 5072) was officially released for shipping by rail to Norfolk, Virginia, on July 1, 1954, its first stop as it made its way to Thule Air Force Base in Greenland. The electric drive system for the vehicle was powered by a 400-hp Allison V-1710 engine modified to run on gasoline. Eventually the Sno-Buggy was returned to Longview and converted into a "Swamp-Buggy" by November 1954, utilizing only four of the Sno-Buggy's original eight wheels.

The unique Series "MA-31" Landing Craft Retriever (LCR) was originally completed by the company in November 1954. Ordered by the U.S. Army, the LCR's main mission was to drive into deep surf and retrieve beached or capsized amphibious vehicles. Only one MA-31 LCR (S/N 5234) was ever built. Image taken on January 25, 1956.

The "XM-1" Corporal missile erector program would go through numerous prototype stages before finally reaching its last approved design configuration as the XM-2 Type II (pictured here in March 1955). LeTourneau only built the erector in the Corporal program. Actual production of the Corporal SSM-A-17 Guided Missile itself was handled by the Firestone Tire and Rubber Company.

Shown in early December 1956 is a pair of "XM-2" Type II Corporal Loader missile launchers (#31 S/N 5371 and #32 S/N 5372) as they make their way out of Longview, Texas, for delivery to Firestone Tire and Rubber Co. in Los Angeles, California.

The Model 6-120 "Tree Crasher" (nicknamed Queenie) was one of the company's first vehicles to feature six electric drive wheels fitted with 120-inch diameter tires. Completed in April 1955, it would eventually ship to Pahokee, Florida, for its first working assignment. After completion, it was returned to the Vicksburg, Mississippi, plant and placed in storage by October 1955. Only one Model 6-120 (S/N 5247) was built. Image taken on May 30, 1955.

Another massive LeTourneau vehicle to feature six wheels fitted with 120-inch tires was the Series A-A (Model CP-1) "Crash Pusher." Built for the U.S. Air Force Strategic Air Command (SAC), the first two units, Fantabulous I (S/N 5264) and Fantabulous II (S/N 5265) were both shipped to Carswell Air Force Base in Fort Worth, Texas, in December 1955, for initial trial testing. Both units would eventually be sent back to Longview for modifications, including the fitting of a "V-plow" blade. After completion of the upgrades, both units were shipped to Davis-Monthan Air Force Base in Arizona, in December 1957. Image taken in June 1958.

The original Model A-6 "Crasher" built in early 1956 featured an elevator-type, rack-and-pinion controlled, bulk material load and carry bucket (pictured). By May of that year it had been fitted with dozing blades at both ends. Image taken on February 16, 1956.

The Model A-6 "Crasher" (S/N 5322) continued to evolve during its field testing program. The third configuration of the same Crasher featured a massive V-plow funnel blade for land clearing operations (pictured). The Model A-6 was powered by a single Cummins NVHBI-1200 engine rated at 400 horsepower. Image taken in August 1957.

Of all the various land-trains developed by the company, none captured the imagination of the public more than the fantastic Model TC-497 Overland Train MkII. Built for the U.S. Army to transport cargo over Arctic, temperate, and tropic zones, it was a true engineering marvel for its time. Power for the electric drive wheels was supplied by four Solar 10MC Gas Turbine engines, each rated at 1,170-hp (4,680-hp total). Image taken in February 1962.

The Overland Train utilized a crew of six under normal deployment. Maximum fuel load was rated at 3,685 gallons, or 7,833 when fitted with the optional extended range fuel tank for the last cargo car. Maximum top speed was 20-miles-per-hour. Image date February 1962.

As originally designed, the Overland Train utilized ten cargo cars plus the power units totaling 54 electric drive wheels. The image taken at Longview in February 1962 only shows nine cargo cars. The tenth was added once the unit was shipped to Yuma, Arizona, for testing. Overall length with all cars in place was approximately 572 feet.

One of the earliest prototype scraper designs to feature the company's electric drive wheel system was the "Rigid A-Tow." The scraper utilized all-wheel drive and steer layout, and was powered by a 450-hp Allison V-1710 engine configured to run on butane fuel. The scraper unit was a modified E50 Carryall. Pictured is A-Tow No.2 (S/N 3596) in August 1951.

The Rigid A-Tow No.2 (S/N 3596) would eventually be converted into the Series DC50 No.2 and issued a new S/N 5054 identification in early 1954. The front rigid mounted Tow nose was scrapped and replaced with a new oscillating axle design. The Allison engine was rebuilt and reused again in the DC50. The unit was then shipped to a Texarkana railway job on April 3, 1954. Of the six A-Tow units produced, No.1 was utilized in the Army Plow program, and numbers 2, 3, 5, and 6 were rebuilt into DC35 and DC50 scrapers. A-Tow No.4 was listed as scrapped as of September 15, 1953. Image date is August 1951.

Similar in design to the Rigid A-Tow was the Rigid C-Tow, also referred to as the Model "E18" DC-Tow #3 (Army 4x4). The Rigid C-Tow was powered by a 218-hp GM 6-71 diesel engine and featured full electric drive and steering wheels. The scraper utilized on the model was a modified E18 Carryall. Pictured is the prototype unit (S/N 2754) minus its front dozing blade, in November 1951.

The original Rigid C-Tow would go through numerous changes and modifications during its development and testing program throughout 1952 and 1953. It would eventually be shipped to Fort Belvoir, Virginia, on June 25, 1954. Pictured in March 1952 is the Rigid C-Tow in its original completed configuration in Longview.

Another early military electric-drive-wheel scraper design was the DC 6x6 Tournatow, also known internally as the C-Tow Dozer (Navy-Tow 6x6). The scraper model utilized a modified E16 Carryall unit, with the tractor unit powered by a GM 6-71 diesel engine. All six wheels were driven and steered. Pictured is the prototype Navy-Tow 6x6 (S/N 2753) at the Longview plant in November 1951.

The Navy Tow 6x6 would eventually be shipped to the Department of the Navy, Fort Belvoir, Virginia, in 1952. It would be returned to the Longview plant on November 24, 1953, and converted into a Series D35 "Army-Tow No.4" and issued a new S/N 5052 identification. Image taken in November 1951.

R.G. LeTourneau continued development of the Navy-Tow 6x6 scraper concept throughout 1953. By mid-year the company had an improved version of the diesel-electric drive scraper ready for U.S. Navy testing in the form of the DC Tractor-T19 (Navy-Tow 6x6). The tractor on this model was equipped with a GM 6-110 diesel engine, and like its predecessor, featured all-wheel drive and steer for the tractor and the scraper. Image date August 1953.

The DC Tractor-T19 (Navy-Tow 6x6) utilized a model T19 scraper unit with powered wheels that also steered. Ultimately four of this type of Navy-Tow 6x6 scrapers were built (S/N 3671/3672/3673/5033), with the first shipping to the Department of the Navy, Port Hueneme, California, on July 3, 1953; the second to Fort Belvoir, Virginia, on October 9, 1953; the third and fourth to the U.S. Atlantic Fleet Construction Battalion Center, Davisville, Rhode Island, on October 31, 1953, and January 21, 1954 respectively. Image date August 1953.

The Series D35 (Army-Tow) was another of the experimental diesel-electric drive scrapers the company would test at a Texarkana, Arkansas, railroad relocation contract conducted under the supervision of the Army Corps of Engineers starting in 1954. The D35 was powered by a 218-hp GM 6-71 diesel engine mounted behind the front drive wheels. All four wheels were powered and had steering capabilities. Scraper model was a T19. Image taken on January 26, 1955, at the Texarkana jobsite.

Looking much like the Series D35, the D50 utilized a slightly modified T19 scraper unit. The pilot D50 was originally equipped with a GM 6-71 diesel (pictured), but it was replaced with a Buda 6DA-844 engine before the scraper shipped to the Texarkana jobsite on September 1, 1954. Shown on April 19, 1954, is the only D50 (Army Tow No.5, S/N 5178) produced.

The Series DC35 was another of the company's experimental scrapers utilized on the Texarkana railroad relocation contract. Power for the DC35 was supplied by a front-mounted, 450-hp Allison V-1710 engine configured to run on butane. The model also featured all-wheel-drive and -steer capabilities. Image taken on January 26, 1955.

Only two DC35 scrapers were built, with one being produced from a previous Rigid A-Tow unit. DC35 No.3 (S/N 5070) was built out of A-Tow No.3 (S/N 3597), and DC35 No.7 (S/N 5074) was a standalone project. Both DC35 models utilized modified versions of the T35 scraper units. Pictured is a DC35 on the Texarkana jobsite on October 28, 1955.

The origins of the Series DC50 scraper project were very similar to that of the DC35 program. In fact all of the DC50 scrapers were assembled out of the remains of disassembled rigid A-Tow units. DC50 No.2 (S/N 5054) utilized A-Tow No.2 (S/N 3596); DC50 No.5 utilized A-Tow No.5 (S/N 3694); and DC50 No.6 utilized A-Tow No.6 (keeping its original S/N 3670 designation). Pictured is DC50 No.6 in Longview on April 29, 1954. It would ship to the Texarkana jobsite on May 7, 1954.

All of the Series DC50 scrapers utilized rebuilt 450-hp Allison V-1710 butane burning engines, and E50 scraper units saved from the donor A-Tow units. Again, like the DC35, the DC50 featured all-wheel-drive and steering. By December 1955, all of the DC35 and DC50 scrapers had been placed in storage at the Vicksburg plant. Pictured is a DC50 in Texarkana on October 28, 1955.

The Series DC (Model C-4) Self-Loading Scraper (also referred to as the 25-Ton Low Gravity Hauler) was the first of the company's diesel electric drive experimental scrapers to feature rack-and-pinion, articulated frame steering. This model was actually built out of Army-Tow No.2 (S/N 5071) and Army-Tow No.4 (S/N 5052, which had been previously been constructed from Navy-Tow 6x6 - S/N 2753). The 25-Ton Low Gravity Hauler would retain the S/N 5052 identification. Image date October 1957.

The 25-Ton Low Gravity Hauler was originally fitted with a 335-hp Cummins NRTO-6 diesel engine. But before leaving the Longview plant, it was fitted with a 600-hp Cummins VT-12-BI unit. Though this model was sometimes referred to as an "Army" scraper, it was never tested by the military. The scraper would eventually ship to a jobsite in San Antonio, Texas, in December 1957. After the unit was shipped back to Longview it was scrapped in December 1958. Image date October 1957.

Another of the company's experimental scrapers to feature an articulated frame steering configuration was the "30-Ton" Low Gravity Hauler (Model C-6). Part of the construction of this model utilized an existing scraper unit (S/N XS-2167-E50) to speed fabrication along. Engine of choice was a single GM 6-110 Turbocharged Diesel. The prototype scraper (also known as the Single-C) would ship to a San Antonio, Texas, jobsite on December 10, 1957. Picture taken at the site on February 10, 1958.

After completion of its testing in San Antonio, the "30-Ton" Low Gravity Hauler was shipped back to the Longview plant and placed in storage. In 1962 the unit was converted to a water tanker and issued a new S/N 6093 identification. A new GM 8V-71 Diesel was also installed at this time. It would then be shipped to a Waco, Texas, jobsite in June 1962. Picture taken in San Antonio on March 28, 1958.

The "80-Ton" Low Gravity Hauler (Model C-10, also referred to as a "Double-C") was a tandem-bowl design driven by ten electric motor powered wheels. It is shown here at work in San Antonio, Texas, on March 27, 1958.

Power for the "80-Ton" Low Gravity Hauler was the tried and true Cummins NVH-BI-1200 diesel engine rated at 400-hp. No serial number was ever issued to this experimental model. Image taken on March 27, 1958.

The largest single bowl experimental scraper design working at the San Antonio, Texas, jobsite was the "70-Ton" Low Gravity Hauler (Model A-6), sometimes called a "Single-A." Engine of choice was the powerful 600-hp Cummins VT-12-M diesel. Image taken in May 1958.

The scraper portion of the "70-Ton" Low Gravity Hauler was fabricated from an existing unit in inventory (XS-2169-E50). The 70-Ton, as well as the 30- and 80-Ton models on the San Antonio site all featured articulated rack-and-pinion steering. All six wheels were driven by electric motors in the wheel assemblies themselves. Image date May 1958.

The design of the "70-Ton" Low Gravity Hauler would greatly influence the next large scraper creations the company would produce, namely the L-70 "Tork Loader" and the L-130 "Electric-Digger." Image date May 1958.

The first piece of earth-moving equipment R.G. LeTourneau introduced after the company's agreement with LeTourneau-Westinghouse not to market this type of machinery had expired in May 1958, was the L-70 (Model A-4) Tork Loader, nicknamed the "Goliath." The L-70 (S/N 5720) would be shipped from the Longview plant, on September 9, 1958, to its world debut at the American Mining Congress exhibition to be held in San Francisco, California. Image taken in August 1958.

Pictured on September 22, 1958, on the opening day of the AMC show in San Francisco is the L-70 "Tork Loader" outside the main convention hall. Engine of choice for the big scraper was the 600-hp Cummins VT-12-BI-600 turbocharged diesel. After the show the L-70 was dismantled and shipped back to the Longview plant. Most of the L-70s components would eventually be utilized in the building of the second L-130 (Model A-8) tandem-bowled digger (S/N 6015) in late 1958.

R.G. LeTourneau built only three of their tandem Series L-130 (Model A-8) Electric-Diggers (No.1 S/N5729; No.2 S/N 6015; and No.3 S/N 6099). By 1962 all would have had their twin Cummins VT-12-BI-600 diesels (1,200-hp) replaced with up to four GM 12V-71 engines (1,680-hp) and rebadged as Series L-140 machines. Pictured is L-130 unit No.1 in October 1958.

The only Series L-140 Electric-Digger to come off the Longview assembly line configured as an L-140 (with four GM 12V-71 diesels) was Model A-8, No.4 (S/N 5989). This unit was actually classified as an L-130, but when its build got underway it was changed to L-140. It would be completed in the summer of 1961. Pictured is the L-140 in December 1961.

In the 1960s R.G. LeTourneau produced a number of Electric-Digger designs featuring a unique single-wheel "tug-pull" drive unit, such as this Series L-67 (Model C-7) from March 1960. Only two tandem L-67s were ever built (S/N 5954 pictured; and S/N 5975). The first unit shown was shipped to LeTourneau of Liberia in December 1960.

The Series L-70 (Model C-12) Electric-Digger (also referred to as a Sand Special Digger) would set the stage for the look of future and even larger triple-bowled scraper models the company would build. In all, seven sets of these triple-bowled L-70s would be shipped into service. Pictured is the first set completed (S/N 6229 issued for front and rear units; S/N 6230 issued to middle unit) in February 1964. It would ship to Modern Machinery Company, Inc., Huntington, Indiana, in March 1964.

Without a doubt, the most popular of the multiple scraper unit models produced by the company in the 1960s was the Series L-90 Electric-Digger. Originally introduced in 1964, this version of the L-90 was capable of handling 57-struck/72-heaped cubic yards, with a maximum payload rating of 90 tons. Pictured is an L-90 (Model A-12) in operation for Harbert Construction Co., Wetumpka, Alabama, on June 1, 1965.

Triple-bowled Series L-90 Electric-Diggers were available in various powered wheel combinations including the Model A-7, A-8, A-11, and A-12, with the 12-wheeled variety being the most popular by far. Pictured on January 1, 1966, was the first of the most powerful version of the L-90 (Model A-12) to be produced by the company, featuring four GM 12V-71 engines rated at 1,900-hp in total. This unit (S/N 6735) would ship to R.A. Heintz Construction Co., Othelo, Washington, by the end of January 1966.

The Pacemaker Series LTV-60 (Model C-6) Electric-Digger featured a telescoping design with a three-V-bucket configuration capable of handling a 60-ton payload. All wheels were driven with power supplied by two GM 12V-71N engines rated at 950-hp. The only LTV-60 (S/N 6237) built (pictured here in October 1963) was shipped to a Morrison-Knudsen job site in Los Banos, California, in January 1964.

One of the largest telescoping V-bucket scraper designs produced by R.G. LeTourneau was the Pacemaker Series LT-110 (Model A-6) Electric-Digger (also known internally as a Quad-Digger). Four GM 12V-71 diesels supplied the power (1,680-hp combined) for all six electric drive wheels. Nominal capacity for the four-V-bucket design was 110 tons (120 tons maximum). The one and only LT-110 (S/N 6112) was originally in the service of Paul Hardeman, Inc., for use on the McGee Bend Reservoir project in Jasper, Texas, in July 1962. It would later be moved to a new jobsite in Waco, Texas, in September 1962. Image date is August 1962.

The massive Series LT-360 Electric-Digger, the world's largest self-propelled scraper ever to be manufactured, was referred to by different designations depending on how many scraper units were being employed at the time. Pictured in one of the Longview plant's "domes" is the lead unit of the LT-360, which was labeled a Series LT-120 (Model A-4) Electric-Digger, on March 1, 1965.

At this stage in the LT-120 Electric-Digger's life it was powered by two 475-hp GM Detroit Diesel 12V-71N engines up front, and two 635-hp 16V-71N units in the rear, totaling 2,220-hp. Tires up front were 98 inches in diameter, while the rear set wore larger 120-inch diameter rubber. Image taken on March 1, 1965.

The LT-120 utilized a telescoping three-bucket scraper design capable of hauling a 72-cu-yd, 120-ton payload. The LT-120 is pictured here on March 9, 1965, in the testing area behind the main Longview plant.

The LT-120 would continue testing throughout the month of March as a standalone unit while the second scraper section was being constructed. It is shown here on March 19, 1965. Note the enclosed cab now fitted to it.

By late April 1965 the second scraper unit was ready for testing with the LT-120, making it now classified as an LT-240 (Model A-6). The front scraper section would forgo one of its rear engines, but then gain two 635-hp 16V-71N diesels from the second unit, bringing the total power output up to 2,855-hp. Capacity in this set-up was 240 tons (144 cubic yards). Image taken on April 29, 1965.

The summer of 1965 would see the LT-360 (Model A-12) in action for the first time. Along with a new middle scraper unit, additional engines and drive wheels were added to the model. There were two 475-hp 12V-71N diesels in the lead "tow-unit" of scraper number one, and an additional six 635-hp 16V-71N engines utilized in the rest of the LT-360. Total power output was a staggering 4,760-hp! This type of output was necessary to power all eight generator sets to supply current to all twelve electric-motor driven wheels, and handle a 360-ton payload. Image taken on August 18, 1965.

The LT-360 (Model A-12) Electric-Digger was officially introduced to the public in August 1965 while working on part of the Interstate Highway 20 project in east Texas, just west of U.S. 259, two miles east of the Sabine River. The actual jobsite bordered LeTourneau property, so the LT-360 was simply driven from the Longview plant to the working area. After completion of the Highway 20 project, it was driven the short distance back to the plant for a complete engineering evaluation. Image taken on August 18, 1965.

Shown on March 2, 1966, is the lead "tow-unit" of the improved LT-360 sticking its nose out of the plant's dome No.2 assembly building, showing off its new set of B.F. Goodrich Tire and Rubber Company's huge 122-inch, 60-inch wide tires (produced by molds engineered by R.G. LeTourneau). Also of note were a pair of new 16-cylinder, 635-hp GM Detroit Diesel 16V-71N engines replacing the previous 12-cylinder 12V-71N units. Larger and more powerful LeTourneau Model A-11 wheel motors (replacing the previous A-3 type) and Wide Double-A drivers were also installed at this time.

Shown on March 3, 1966, is the LT-240 portion of the improved LT-360 parked behind the Longview plant, in the testing area referred to as "Red Hill."

The LT-240 section of the LT-360 would make several test loads during the day behind the Longview plant as the third scraper unit was readied. Image taken on March 3, 1966.

On March 4, 1966, the Series LT-360 (Model A-8) Electric-Digger was seen for the first time fully assembled with all three scraper units ready for inspection. It was simply the world's largest and most powerful scraper model ever assembled.

This famous publicity shot from March 4, 1966, showing R. G. LeTourneau himself and his VW Beetle posing next to the improved LT-360, really sets the scale as to just how large the scraper was in real life.

Modifications to the original LT-360 included a revamped drive wheel configuration layout (now eight wheels instead of twelve), larger tires, wheels, and A-series drivers, plus more horsepower thanks to an engine upgrade. Now all eight generator sets were powered by eight 635-hp GM Detroit Diesel 16V-71N engines, producing an astonishing 5,080-hp combined! Image date is March 4, 1966.

Capacity of the LT-360 (Model A-8) was approximately 216 cubic yards, with a maximum payload capacity of 360 tons. Each of the scrapers lead cutting edges measured 14 feet, 2 inches in width. 16mm color film of the LT-360 taken in action at the time showed that the entire unit was surprisingly fast during its loading sequence, and that a single operator had no problem putting it through its paces, despite the scrapers approximate 200-foot length and 19-foot width. Image date March 4, 1966.

The tires on the upgraded LT-360 were massive 60.00x68, 72-ply series, measuring 122 inches in diameter. Even though the tires said R.G. LeTourneau on the sidewalls, they were actually cast by B.F. Goodrich in molds engineered by LeTourneau. Image date March 4, 1966.

This view really illustrates just how long the LT-360 was in comparison to a full-size automobile. But the entire unit was controlled by a single operator from the control cab mounted to the left rear of scraper unit one. Image date March 4, 1966.

Sadly, the LT-360 would never be placed into service. In fact, by May 1966 many of its components were already being removed to be utilized on other R.G. LeTourneau Electric-Digger prototypes (no serial numbers were ever assigned to the LT-360 program). The LT-360 was simply too big for its own good. The LT-360 still holds the record of the world's largest scraper ever built to this very day. Image taken on March 10, 1966.

To get an ultra-high capacity Electric-Digger that was of a more "manageable" size relatively speaking, the company introduced an experimental model called the Series LT-300 (Model A-6). This scraper design's tandem bowl design measured 148 feet in length, compared to the LT-360's 200-foot measurement. Pictured on June 3, 1966, is the front control unit of the massive scraper. As a standalone unit it was identified as a Series LT-180 (Model A-4).

This image from mid-June 1966 shows the completed LT-300 (Model A-6) testing behind the main plant near the factories domes No.4 and 5. The LT-300 was powered by seven 635-hp GM Detroit Diesel 16V-71N engines (four on the front scraper and three on the rear unit), totaling 4,445-hp. The Model also utilized the same sized tires, wheels, and wheel motors and drivers as the LT-360.

The LT-300 Electric-Digger was capable of handling a 300-ton payload (180 tons in the front unit and 120 at the rear). The rear scraper unit on the LT-300 was in fact the third scraper unit utilized from the upgraded LT-360. It had been removed from that model in late April 1966. Shown on August 19, 1966, is the only LT-300 to be built. The LT-300 never left the Longview facilities.

Another benefactor of components from the LT-360 program was the experimental Series 5-B-90 (Model A-4) Electric-Digger. The 5-B-90 was a telescoping five-bucket design rated at 90 tons in capacity. The front "tow-unit" of this model was saved from the LT-360, but with smaller 98-inch tires and C-9 wheel motors with A-3 drivers installed. Engines were the same GM DD 16V-71N units rated at 1,270-hp combined. In March 1967 a powered trailer fabricated out of an old Series LTV-60 was attached that added another 60 tons to the payload rating. Image taken on September 28, 1966.

The last of the ultra-large capacity Electric-Diggers to be designed and built by the company was the Series 6-B-150 (Model A-4). Its telescoping six-bucket scraper design had a nominal payload rating of 150 tons, with a maximum rating of 180 tons. Image taken on June 2, 1967.

The 6-B-150 was powered by three 635-hp GM DD 16V-71N engines rated at 1,905-hp combined. The model also utilized 122-inch diameter tires, and powerful LeTourneau A-11 wheel motors and huge Double-A drivers, the same as installed on the previous giant LT-360 and LT-300 Electric-Digger creations. Steering was through the rear wheels only. Image taken on June 2, 1967.

It soon became evident through early field testing that the 6-B-150's frame needed further reinforcement. Shown on July 13, 1967, is the 6-B-150 with beefed-up "X" side bracing to help reduce flexing of the main frame. A planned 150-ton capacity powered trailing unit was also on the drawing board to expand the design to a Series 12-B-300, but it died along with the original 6-B-150 program when the company terminated the projects due to the firm's financial difficulties at the time.

The Pacemaker Series LTC-75 (Model A-42) Electric-Digger was one of the company's more conventional looking scraper models. Originally built in April 1963, it was powered by a pair of 475-hp GM 12V-71N diesel engines rated at 950-hp combined. Pictured in June 1963 is the first LTC-75 manufactured (S/N 6163) working a coal stockpile at the Tennessee Valley Authority's (TVA) Shawnee Steam Plant, in Paducah, Kentucky..

The Pacemaker Series LTC-75 was rated at 75-struck/91-heaped cubic yards, with an overall payload capacity of 65 tons. By 1966 the designation of the LTC-75 was changed to simply LC-75. Total number of LTC/LC-75 Electric-Digger coal haulers built was only five. Pictured at the Longview plant on September 2, 1965, is LTC-75 No.4 (S/N 6557) ready for shipping to Tennessee Valley Authority's Bull Run Steam Plant in Tennessee.

The last of the company's Electric-Digger scraper models was the Series SC-35B (Model C-4). The SC-35B coal scraper utilized a single 475-hp GM DD 12V-71N engine, and was rated at 32-struck/39-heaped cubic yards. Payload capacity was 30 tons. Only two SC-35B coal scrapers would be built (S/N 7167/7186) and both would be sold to Dofasco, in Hamilton, Ontario, Canada. Pictured in May 1968 is the first SC-35B completed by the company.

Here a Pacemaker Series K-100 (Model A-4) Electric Tractor simply dwarfs a Caterpillar DW-20 scraper as it push-loads it. Power for the K-100 was supplied by a single 600-hp Cummins VT-12-BI-600 diesel engine. Weight of the K-100 equipped with the 15-foot wide dozing blade was 106,280 pounds. Image date is July 1959.

The largest of all the "tug" dozers produced by R.G. LeTourneau was the Pacemaker Series K-205 (Model A-5) Tractor. All five of the wheels were driven giving the K-205 excellent traction for dozing and push-loading applications. Shown in March 1961 is the K-205 (S/N 6084).

The Pacemaker Series K-205 was powered by three 420-hp Cummins V-12-525 diesel engines, totaling 1,260-hp. All five wheels were driven by electric wheel motors just like all of the rest of the company's vehicles during this time period. Image date March 1961.

The K-205 original tires were 89 inches in diameter. In 1962 the tractor would be fitted with larger 98-inch diameter rubber for increased traction while push-loading scrapers. Image date March 1961.

Shown at work in May 1962 is the K-205 in its first month of operation for W.A. Smith Contracting Company, Inc., at the Hubbard Creek Dam site, Breckenridge, Texas, push-loading a Euclid SS-40 scraper. This model of Euclid scraper was a large machine in its own right, but the K-205 is definitely up to the task both size- and power-wise.

The Pacemaker Series K-205 utilized a blade that was 20 feet wide and 4 feet, 6 inches high. Overall length was listed at 50 feet, 3 inches, with an operating weight of 160 tons. This view of the K-205 from December 1964 clearly shows the redesigned front rack-and-pinion blade control mounting (which was first seen on the tractor in October 1963). Only one K-205 was ever produced.

The Longview-built experimental Series SL-20 Electric-Tractor models were built in quite a few unique configurations. The more conventional of these was the SL-20 (Model C-4) shown testing here on April 2, 1965. The rack-and-pinion controlled front-end loader carried a 20-ton capacity bucket and was powered by a single GM 475-hp 12V-71N diesel engine. Only one of this type of SL-20 was produced (S/N 6528).

Another variation on the front-end loader SL-20 theme was this model from late 1965, which featured a novel "roll-out" bucket design. Engine installed on this loader was a Dorman Model 60TCA diesel. The loader would ship to Lone Star Steel in Texas for field evaluation testing in mid-December 1965. Only one with this bucket configuration was ever built (S/N 6730). Image taken on December 9, 1965, in Longview.

Shown on March 17, 1966, is a modified version of the original SL-20 (S/N 6528) Electric-Tractor, now equipped with an additional cab and dozer blade on the engine chassis portion of the loader. The dual-cab SL-20-D-L (Model C-4) Electric-Tractor was shipped in April 1966 to the Quebec Iron & Titanium Corp., located in Sorel, Quebec, Canada. Image taken on March 17, 1966.

One of the more unusual front-end loaders developed by the company was the Series SL-20-R (Model C-4) Revolving Bucket Tractor. This model featured a single 475-hp GM 12V-71N engine, a rear-mounted dozing blade, and a front-end loader attachment capable of a full 180-degree rotation. The only SL-20-R (S/N 6765) was shipped to Tournavista in Peru, in April 1966. Pictured at the Longview plant in February 1966 is the SL-20-R loading the TR-30 Electric Rear Dump.

Next size up from the SL-20 in the company's front-end loader program was the Series SL-30 (Model A-4) Electric-Tractor. First unveiled in February 1964, the SL-30 was rated as a 30-ton capacity loader. Power was supplied by two rear-mounted, 475-hp GM 12V-71N diesels (950-hp total output). The first SL-30 (S/N 6307) is shown here in June 1964 ready for delivery to Buckeye State Machinery, Inc., in Toledo, Ohio, for field testing at Ohio Lime.

The second SL-30 produced (S/N 6311 pictured) featured a relocated operator's cab to the design of the loader line. In all there were only three SL-30 loaders manufactured (S/N 6307/6311/6353) and all were converted to full Series SL-40 specifications. The quickest way to tell the difference between the two models is that the SL-30's engines are mounted transverse style, while the SL-40's are side-by-side in line with the chassis. Image date is July 1964.

Shown on March 3, 1965, at the Longview plant is a full spec Series SL-40 (Model A-4) Electric-Tractor, wearing an optional larger tire and wheel package. Capacity of the SL-40 loaders bucket was 19 cubic yards, with a maximum payload weight of 40 tons.

The 950-hp Series SL-40 would make an appearance at the 1965 American Mining Congress exhibition in Las Vegas, Nevada, in full production dress. Engine type and power output of the SL-40 was the same as the SL-30 it replaced.

All SL-40 Electric-Tractor loaders were a rather rare machine to begin with, but one was more so than the rest. Shown testing at the Longview plant on March 9, 1967, was a one-of-a-kind SL-40 modification featuring new front lifting arm assembly and rack-and-pinion configuration. In all there were a total of ten SL-40 loaders built, three of which were rebuilt SL-30 units turned into SL-40s.

Another view of the right side of the SL-40 featuring the new front-end loading assembly. Tires on the model shown were 98 inches in diameter. Image taken on March 9, 1967.

This close-up detail of the new simplified rack-and-pinion lift-arm assembly on the modified SL-40 clearly shows the larger housing necessary for the bigger electrical hoist motors and huge drivers for the rack, which is now strait instead of curved as in all the other examples of this model. Image taken on March 9, 1967.

R.G. LeTourneau produced a small number of rubber-tired dozers in the 1960s that were as modern looking (except for their use of rack-and-pinion steering and blade controls) as anything else available in the marketplace at the time. One of those was the T-450 series. Originally introduced in September 1965 as the Series K-54 (Model CA-4), it was re-designated the Model T-450-A in January 1966. It would be upgraded into a T-450-B "LeTro-Dozer" series in April 1970. Total number built of all models was only 15 units (four K-54s; eight T-450As; and three T-450-Bs). Image taken in May 1968.

The largest of the early R.G. LeTourneau rubber-tired electric-drive dozers was the Model T-600-A. Originally built in June 1967 as the Series K-600-A (Model A-4), it was re-classified as the Model T-600-A in September 1967. Only four of the big 635-hp dozers were eventually built (three T-600-As; and one T-600-B). Pictured in October 1967 is the prototype unit (S/N 6888) during field trial testing with Fort Myers Construction Co. on their Cape Coral project in Florida.

The company's "Power-Packer" diesel-electric-drive sheepsfoot rollers could actually be considered a moderate success for the company with approximately 35 produced of all model types. The largest of these was the Series M 60-55 (Model 5x5) first produced in September 1964. The M 60-55 was powered by two 475-hp GM Detroit Diesel 12V-71N engines (950-hp total) mounted behind the operator's cab. Pictured in May 1965 is the only M 60-55 Power-Packer built (S/N 6320), originally produced with two engines and not a modified Series M 50-55.

The company tried several times to produce a diesel-electric drive, off-highway quarry and mining hauler model for the marketplace. One of those attempts was the Pacemaker Series TR-30 (Model C-2x4) Electric Rear Dump introduced in early 1963. The 37.5-ton capacity hauler was powered by a single 420-hp GM 12V-71 diesel engine that supplied power to the front wheels only. Only one TR-30 (S/N 6166) was ever built. Image taken in March 1963.

R.G. LeTourneau's first serious attempt at producing a marketable off-highway mining hauler was the Series TR-60 (Model A-4) "Trolly-Dump." The diesel-electric drive truck's main power was supplied by a 600-volt DC electric overhead trolley line which gave the TR-60 a maximum power output of 1,600-hp. When off the overhead line, a single 335-hp Cummins NRTO-6 turbo-diesel was utilized for maneuvering power. Capacity was listed at 65 tons in this power configuration. The TR-60 was completed in July 1959, and it was shipped to Anaconda Company's Berkeley Pit in Butte, Montana, in August of that year. Image date August 1960.

It soon became evident that the TR-60 needed more power when not connected to the overhead power supply. A new front "tow-unit" was constructed, equipped with two 335-hp Cummins NRTO-6 diesels, giving the unit 670-hp of maneuvering power when off the trolley lines. A newly redesigned operator's cab was also installed at this time. The new engines allowed capacity to be increased to 75 tons. Only one TR-60 (S/N 5793) was ever built. Pictured in February 1961 is the hauler with the new engine and cab configuration.

Another of the company's experimental hauler programs was the 100-ton side-dump. This hauler was originally conceived as a four-wheel, diesel-electric drive, front-engined side-dump powered by two 475-hp GM Detroit Diesel 12V-71N engines (950-hp combined). Pictured is the pilot Series TS-100 (Model A-4) Electric Side-Dump on May 17, 1965.

After months of testing LeTourneau engineers decided that the TS-100 needed more powered wheels to address traction and stability concerns. Shown on March 29, 1967, is the reconfigured Series TS-100 (Model A-6) now equipped with six-wheel electric drive.

The TS-100 (Model A-6) was actually the original four-wheeled unit heavily modified and reengineered to a six-wheel chassis layout. Capacity remained unchanged at 100 tons. Image taken on March 29, 1967.

Over the next several months the TS-100 (Model A-6) would undergo still more changes in its overall layout. Pictured on June 21, 1967, is the "third" generation of the TS-100, now with the engines moved to the rear.

The two original GM 12V-71N engines were still utilized, but now they were moved to the rear of the TS-100 for better weight distribution and traction. But the fate of the TS-100 fared no better than the previous TR-60 and TR-30 hauler programs. Only one TS-100 (S/N 6869) was ever produced (all three configurations were made on the same chassis). It was listed as being shipped to N'Changa Consolidated Copper Mines, Ltd., in N'Changa, Zambia, in July 1967. Image date is June 21, 1967.

Numerous self-propelled water tankers were produced by the company that were diesel-electric drive in nature. One of the largest of these was the Series W-12 (Model A-6) Water Wagon. The big W-12 was capable of hauling 12,000 gallons of water on six electric-motor-powered wheels fitted with 120-inch diameter tires. Power for the water wagon was supplied by a single 635-hp GM 16V-71N diesel engine. Image taken on October 5, 1965.

The W-12 Water Wagon would ship to Kennecott Copper Corporation's Utah Copper Division's Bingham Canyon operations in Utah, in November 1965. In early 1966 the W-12 was fitted with additional high-floatation tires (six up front and two per wheel assembly at the rear) for a total of 14 tires. Only one W-12 (S/N 6591) was ever delivered into service. Image taken on April 20, 1966.

R.G. LeTourneau constructed many unusual and one-of-a-kind vehicles utilizing old military surplus tank chassis. Donor models included the M4A3 HVSS "Sherman" and the 38-ton M6 High Speed Tractor. Pictured in April 1961 Sherman tank chassis converted into a dozer utilizing a diesel-electric drive system.

During the 1960s the company produced a number of three-wheeled, diesel-electric drive tractors that featured a rear powered "tug" unit. One of those was the Series K-53 (Model CA-3) Swamp-Dozer, better known as the "Billy-Goat." As built in June 1964, the Billy-Goat was powered by a single 420-hp GM 12V-71N engine, and rode on five 120-inch high-flotation tires. It was also equipped with a front-mounted 25-foot, 10-inch wide brush cutter. Image taken in June 1964.

Shown in September 1964 is the Series K-53 (Billy-Goat) Swamp-Dozer fitted with a more powerful 635-hp GM 16V-71N diesel engine, and two additional 120-inch diameter tires on the front two drive wheel assemblies.

By late 1965 the Billy-Goat's high-flotation tires were replaced with 98-inch diameter steel drive wheels for working in harsh, dry conditions. Only one K-53 Swamp-Dozer (S/N 6325) was ever produced. Image taken on November 3, 1965.

R.G. LeTourneau's Log Stackers proved to be one of the company's most important product lines during the late 1950s and all of the 1960s. As a volume leader in sales, they were the models that kept the company going as they continued to develop new earth-moving equipment lines. Pictured on July 18, 1955, is the first diesel-electric drive pilot Series "F" Log Stacker (Model C-4) produced (S/N 5252).

The Series "H" (Model 6-16) Disc-Plow was originally built for large land clearing applications. Key to the models design were its massive disc blades, each measuring just over 6 feet in diameter. Power for the electric drive system was supplied by a single front-mounted Buda 8 DAS-1125 diesel. The Disc-Plow would eventually be shipped to Buckeye Cellulose Corp., in Foley, Florida, in July 1956. Image taken on August 15, 1955.

In the fall of 1956 the Series "H" Disc-Plow was fitted with a set of steel drive wheels on the front only. The rear axles would retain the use of normal tires. Image date October 1956.

The Series "H" Disc-Plow was eventually shipped back to Longview for a complete rebuild. The original Buda diesel was removed and replaced with a 420-hp GM 12V-71 engine. Pictured on August 6, 1965, is the Series "H" Disc-Plow (S/N 5335) equipped with the new engine module. Only one Disc-Plow was ever built.

R.G. LeTourneau produced a number of large tree crushers (or jungle crushers) starting in the mid-1950s through all of the 1960s. Some were configured with three crushing steel wheels, while others utilized the two roller design, such as this 110-ton Series "G" Tree Crusher from December 1956.

Pictured at the Vicksburg, Mississippi, plant in October 1958 are two models that were produced there. The smaller one on the right is the Single A-12 Tree Crusher (S/N 5725), more commonly referred to as the Series G-50. On the left is the huge Series G-175 (S/N 6085), the largest crusher the company ever produced. The G-50 utilized 12-foot wide rollers, while the G-175 was fitted with super-wide 30-foot units.

Many of the high-floatation three "star-wheeled" configured tree crushers were designed for use in swamp-type land clearing applications. Pictured testing on March 17, 1966, is the Model 40x40 Tree Crusher (S/N 6766). Weighing in at 176,340 pounds, the 40x40 utilized front drum-wheels that were 16 feet wide each, with the rear drum measuring 9 feet across.

This image of the Model 40x40 from April 8, 1966, shows the relocated operator's cab that has been moved to the rear of the unit for better visibility and safety. Power for the unit was supplied by a single 635-hp GM Detroit Diesel 16V-71N engine. The 40x40 was originally scheduled to be shipped to the Tournavista compound in Peru, in late April, but was rerouted to the Toledo Bend Reservoir project in Many, Louisiana, in June 1966 instead.

Another "star-wheeled" crusher design was the Model S-60 Plow Piling Crusher. This crusher originally started testing in October 1965 equipped with a massive front-mounted push-beam. This was eventually replaced with a large root-rake blade in 1966. The operator's cab was also relocated at this time for better visibility. Power for the 160,000-pound S-60 was supplied by a single 475-hp GM 12V-71N engine. The only unit built (S/N 6769) would ship to RGL Corp., in Orange, Texas, in May 1966. Image taken on July 29, 1966.

Introduced in early 1967, the Model 35x35 (also known as the Series 35) was one of the heaviest of the three drum-wheeled designs, tipping the scales at 214,965 pounds. The 35x35 utilized three 12-foot wide octagon scalloped-sided steel drum-wheels, all powered by a single 635-hp GM 16V-71N engine. The large front-mounted push beam measured 35 feet in width. The only 35x35 built (S/N 6826) was shipped to Toledo Bend, Many, Louisiana, in February 1967. Image taken on February 3, 1967.

The Model 34x30 "Transphibian Tactical Crusher" was built for use by the U.S. Army for land clearing duties in Vietnam. Two sets of wheels were provided – one foam-filled, hexagon scalloped sided set, and one drum-type set fitted with raised chevron-type blade grousers. Picture on April 11, 1967, testing in Longview, is the hexagon star-drum type which measured 14 feet wide, with a 12-foot diameter each.

Pictured on May 8, 1967, at the Tally Bottom testing area not far from the main Longview plant, is one of the 34x30 Tactical Crushers fitted with the chevron-bladed drum wheels. The frame of the 34x30 has also been modified and strengthened at this point in the testing program of the unit. Power for the crusher was supplied by a single 475-hp GM 12V-71N engine. Maximum operating weight was listed at 202,215 when equipped with the hexagon star-drums.

Only two Model 34x30 Transphibian Tactical Crushers (S/N 6874/6875) were ever built, and both were deployed to South Vietnam. By late 1968 the U.S. Army deemed the Tactical Crushers too impractical for operations in Southeast Asia and they were returned to the United States and resold. Pictured in July 1967 is one of the crushers operating in South Vietnam.

The first of the three-wheeled crusher models to be introduced by the company was the Series G-40 in the summer of 1960. The G-40 was equipped with a single 420-hp GM 12V-71 engine, which supplied power to all three electric drive wheels. Operating weight was approximately 84,000 pounds. Shown on the Longview assembly line in May 1960 is the pilot G-40 (S/N 5836). Total number of G-40 crushers built was only 10 units.

To ease in the transport of the large three-wheeled crushers, models like the Series G-60 (pictured) were designed with a front-end drive assembly that could be detached from the main chassis and transported separately. Here a G-60 is being readied for transport at a Georgia-Kraft Company operation in north Georgia, in September 1967.

A heavier-duty model than the G-40 was the Series G-80 Tree Crusher introduced in 1961. The G-80 utilized more powerful A-series wheel motors, while the G-40/55/60 series crushers were equipped with C-series units. Pictured in October 1961 is the pilot G-80 (S/N 6005) which was powered by a two 420-hp GM 12V-71 engines (840-hp combined).

Shown in January 1962 is the second Series G-80 Tree Crusher built (S/N 6047). This unit featured a reinforced front push-beam and heavier-duty operator's cab. Power for this G-80 was supplied by one GM 12V-71 and one GM 8V-71 engine (720-hp combined). Overall working weight of the G-80 was listed at 60 tons. Only three Series G-80 crushers were ever built.

In 1968 the company replaced the Series G-80 Tree Crusher with the newly redesigned Model G-80B. The G-80B featured new knife-blade-like grouser-equipped drum rollers that were 10 feet wide and 7 feet in diameter. Power was provided by a single 635-hp GM Detroit Diesel 16V-71N engine. Overall working weight was listed at 70 tons. Only four G-80B crushers were produced. Image taken on January 5, 1968.

R.G. LeTourneau introduced a smaller crusher in 1970 in the form of the Model 3523 LeTro-Crusher, which was designed for clearing applications involving smaller tracks of land. Its diesel-electric drive system was powered by a single 318-hp GM Detroit Diesel 8V-71N engine. Overall working weight was approximately 38 tons. Pictured on January 23, 1970, is the pilot Model 3523 (S/N 1027). It would ship to Hudson Pulp & Paper in Palatke, Florida, in March 1970. The total number of 3,523 LeTro-Crushers manufactured was 35 units when production ended in mid-1981.

LeTourneau often supplied their components to other manufacturers, such as this unidentified builder of a self-propelled tunnel boring machine from the early 1970s. In this application LeTourneau supplied the tires, rims, and drivers, along with engineering assistance.

Another unique engineering challenge presented to R.G. LeTourneau in the late 1960s was the LC-17 Propulsion Drive Truck. Built for use by NASA at Cape Kennedy (now referred to as Kennedy Space Center), they are the main drive trucks for the Mobile Service Tower (MST) utilized at the Space Launch Complex (SLC) 17, pad A or B. LeTourneau delivered the first four LC-17 trucks in October 1967, with the next four for the second tower (classified as LC-17B trucks) in August 1969. The trucks themselves were assembled at the Vicksburg plant, with the main M-G electri-

cal propulsion modules (JB-9700) that powered them being supplied by the Longview facilities. The two MST utilizing the eight LeTourneau LC-17 trucks are still in service today providing support for the Delta II launch vehicle at SLC 17. Image date is October 1967.

In the spring of 1965 R.G. LeTourneau started a program on the development on a diesel-electric drive semi-tractor unit. Utilizing a GMC "Crackerbox" cab-over as a starting point, engineers installed a GM diesel powered generator set behind the main cab which supplied power to an electric motor attached to the main transmission case. Two prototypes were built, but only the second unit (pictured S/N 6778) was licensed for legal highway road testing use. Image taken on April 12, 1966.

The Series MDS-200 (Model RD-230) Mobile Gantry Crane was one of the company's larger dock-side lifting cranes. Constructed for Niagara Frontier Port Authority for use at the Port of Buffalo in New York, the MDS-200 had a standard rated work capacity of 50 tons at 40 feet, and a maximum lift capacity of 75 tons at 28 feet on its 100-foot boom. The Longview-built mobile crane was completed in May 1967 and dedicated into service by Niagara on July 27, 1967. Pictured is the only MDS-200 produced (S/N 6843) in July 1967.

Similar in design to other LeTourneau mobile gantry cranes, the Model MGR-M 140 LeTric Operated Crane featured a unique multi-wheeled, four bogie, 32-wheel undercarriage. Maximum lift capacity was rated at 50 tons at 22.5 feet on its 85-foot boom. The crane was originally identified as a Series MDS-140, but was changed to the Model MGR-M 140 just before final delivery to a C&O Railway port facility in Norfolk, Virginia, in January 1968. It would go into full service in July of that year. Only one MGR-M 140 (S/N 6914) was ever built, and it is pictured here at the Vicksburg plant on November 20, 1967.

The largest of all of the company's dock-side cranes was the eight-wheeled Model MGR-500 Mobile Gantry Revolving Crane. The crane had a maximum lifting capacity of 75 tons at 28 feet on its 120-foot boom, and had an overall working weight of 708,500 pounds. When the first unit (S/N 6997) was released for delivery to the Port of Philadelphia in November 1968, it was identified as the MGR-500. In 1970 the crane's nomenclature was changed to the Model GC-500 LeTro-Crane. Only two Longview-built MGR/GC-500 cranes were built, with the second unit (S/N 7175) shipping in late 1971 to the Long Beach Naval Shipyard in California. Pictured is the first MGR-500 assembled on November 7, 1968.

Two of the largest RD-Series Electric Cranes built by R.G. LeTourneau were the Series RD-1600 and the RD-1200 (pictured). The RD-1200 was barge-mounted to LeTourneau Floating Crane Barge Hull No. 15, and was christened the "Jirafa." The RD-1200 had a maximum lifting capacity of 145 tons, with a boom length of 274 feet. Built at the company's Vicksburg, Mississippi, yards, it was officially handed over for service in August 1960. Image taken in Vicksburg in July 1960.

LeTourneau Hull No. 15 equipped with the RD-1200 was purchased by Consorcio Puente Maracaibo, in Caracas, Venezuela, for bridge work on Lake Maracaibo. This rare close-up from July 1960 of the RD-1200 on the "Jirafa" showcases the lower works of the barge-mounted crane.

The largest of all the "RD" cranes was the Series RD-1600 mounted to self-elevating LeTourneau Mobile Island (also referred to as a Floating Island Crane) Hull No. 10 "Elephante." The RD-1600 had a maximum lifting capacity of 250 tons on its 200-foot boom. It would be delivered into service in November 1959, and it is pictured here in January 1960 working on the same Lake Maracaibo bridge contract as the "Jirafa." Both the Elephante and Jirafa were owned by the same company in Venezuela. The Elephante was eventually sold and placed into the service of Coral Drilling and renamed the "Little Bob." Unfortunately, Hull No. 10 "Little Bob" was destroyed in a fire off the coast of Louisiana in August 1968.

Chapter 3 – The Modern Era

The modern era of LeTourneau equipment history starts off in 1968 with the building of the company's first diesel-electric drive front-end loader to feature hydraulic controls, the XL-1, soon to be known as the L-700 "LeTric-Loader." It would be officially introduced to the public the following year. The look and mechanical layout of the L-700 would establish LeTourneau once and for all as a legitimate player in the quarry and mining marketplace, at least on the upper end in size and performance.

The old rack-and pinion-controlled designs of the past were considered by most in the heavy-equipment industry as oddities that made little impact in the market. The exception to this was the company's log stackers. Introduced in 1955, they were the one product line that helped keep LeTourneau solvent during the late 1950s and the 1960s as the new earth-moving models were being developed by the company at great expense. The forestry equipment designs would eventually take a backseat to the new hydraulic-controlled mining loaders as they started to enter the marketplace in greater numbers. Today, the log stackers still survive as a limited production product line, intact with the old rack-and-pinion controls. Antiquated? Yes. But ultra-reliable and well suited for the working applications they were designed for.

During this time period LeTourneau as a corporate identity has operated under a handful of names, such as R.G. LeTourneau, Inc., Marathon LeTourneau, LeTourneau, Inc., and LeTourneau Technologies, Inc. (LTI). On May 16, 2011 Joy Global, Inc. of Milwaukee, Wisconsin, announced it had purchased LTI from its corporate parent, Rowan Companies. This transaction would be completed on June 22, 2011. The purchase included the Longview, Texas, and Vicksburg, Mississippi, operations, with the Vicksburg plant (and all marine and off-shore drilling products) to be quickly sold off to Cameron International Corp., in a deal that would close on October 24, 2011. On June 1, 2013, the Longview facilities would be officially renamed Joy Global Longview Operations LLC.

As for the LeTourneau mining product lines after the acquisition by Joy Global, they were sold under the P&H brand name as a P&H LeTourneau-Series machine. But in July 2014, Joy Global officially announced that it was removing the "LeTourneau" name from the mining product lines. Today the giant loaders produced in Longview are marketed by Joy Global as "P&H Generation 2" wheel loaders, with the individual machines themselves being referred to as a P&H L-1850, P&H L-2350, etc. As a tribute to its heritage (and as a way to protect and maintain the trademark), a small "LeTourneau" plate is now fitted inside the cab, usually on the right side of the dash. At the time of this writing Joy Global's LeTourneau Forestry Products will retain the LeTourneau name on the log stackers produced in Troutdale, Oregon, for the foreseeable future.

R.G. LeTourneau's first front-end, rubber-tired, diesel-electric drive wheel loader, built from the ground up incorporating hydraulic control systems was the "XL-1" LeTric-Loader. Image taken on March 7, 1968.

The pilot XL-1 LeTric-Loader was as modern looking as any front-end loader being produced by the competition. For the company it would be one of the most important machine designs in its modern history. Image date March 7, 1968.

The XL-1 loader's diesel-electric drivetrain was powered by a single 16-cylinder, 700-hp GM Detroit Diesel 16V-71T-N75 engine. The drivetrain's generator supplied power to A-series wheel motors in the front and C-series motors at the rear. By November 1968 all wheel motors installed would be large A-series units. The company referred to this drive-system at the time as LeTro-matic DC Drive. Image date March 7, 1968.

As modern as the XL-1 was engineering-wise for the company, they still featured an antiquated rack-and-pinion steering mechanism for the loader's articulated steering controls. Thankfully, this would be replaced with a more modern pivot-ball/hydraulic cylinder arrangement in early 1969. Image date March 7, 1968.

Capacity for the XL-1 LeTric-Loader was 15 cubic-yards, with a maximum bucket capacity of 45,000 pounds. Overall working weight was listed at 175,000 pounds (87.5 tons). Image date March 1968.

The new XL-1 (now officially designated the L-700) was officially introduced to potential customers during a special event held on February 19, 1969, at the company's Longview, Texas, facilities. Pictured is the pilot L-700 parked next to its little brother, the L-500 (which also made its first public showing), at the Gregg County Airport jobsite demonstration area, located just a few miles from the main Longview plant.

Shown in Longview on April 12, 1970, is the pilot L-700 (S/N L-700-1027) ready for shipping to Utah Construction & Mining's Navajo Mine, located near Farmington, New Mexico. The pilot L-700 (now referred to as a LeTro-Loader) at this point in its life is now fitted with a 725-hp Waukesha L1616DSI diesel engine, and a 23.5-cu-yd coal loading bucket.

The pilot L-700 would go through some very comprehensive upgrade programs during its working life. In 1972 the loader was fitted with partial solid-state controls, making it the first L-700A. In 1973 it was rebuilt with full solid-state controls with solid-state converters/controllers, making it also the first L-700SS (also considered the first prototype L-800). Three model designations, all the same base machine (S/N L-700-1027). Pictured is the prototype L-700 operating at the Navajo Mine on May 4, 1970.

Shown loading a 100-ton capacity Unit Rig Lectra Haul M-100 at Gaspe Bay Cooper in Quebec, Canada, in June 1975, is an L-700A LeTro-Loader.

This Marathon Le-Tourneau L-700A LeTro-Loader is loading a bottom-dump coal hauler with a 23.5-cu-yd coal loading bucket at Utah Construction & Mining's Navajo Mine, located near Farmington, New Mexico, in May 1975. The mine operated up to five L-700/L-700A loaders during the 1970s.

The L-700 LeTro-Loader was a solid success for LeTourneau and would firmly establish the new product line in the mining marketplace for the company with a total of 76 units shipped into service worldwide. The last L-700A shipped from the Longview plant in March 1977. Pictured in 1974 is a full-spec L-700A model ready for delivery.

The XL-1 (L-700) smaller brother was the XL-2 (L-500) LeTric-Loader. The design program for the XL-2 program started in mid-1968, with the first (and only) unit completed by February 1969. Designated the L-500, it is shown here at the Longview plant on February 26, 1969.

The L-500 LeTric-Loader's diesel-electric drivetrain was powered by a single 500-hp, 12-cylinder Cummins V-1710-C-500 engine. Bucket capacity was listed at 10 cu-yds, with a 30,000-pound payload rating. Image date February 26, 1969.

After initial testing around the main LeTourneau plant, the L-500 was shipped to Lone Star Steel in Texas, located north of Longview, for field trails. But financial difficulties at LeTourneau would cause the L-500 program to be shut down. Pictured on November 18, 1969, is the only L-500 LeTric-Loader produced in operation at Lone Star Steel.

Designed at the same time as the L-500 (XL-2) was its sister machine, the D-450B (XT-1) LeTric-Dozer. The diesel-electric drive dozer carried a 14-foot, 3-inch wide blade that measured 60 inches in height. Overall working weight with hydroflation was 105,000 pounds. Image date February 26, 1969.

Power for the D-450B was supplied by a 530-hp GM Detroit Diesel 12V-71N engine. And as in all previous loader models, all four wheels were driven by internal electric wheel motors. Pictured on February 19, 1969, is the only D-450B LeTric-Dozer built during the Gregg County Airport demonstration that took place not far from the main Longview plant. The LeTric-Dozer program was cancelled (along with the L-500 LeTric-Loader) due to financial cutbacks at the company.

Introduced in 1976, The Marathon LeTourneau L-600 LeTro-Loader was a 10-cu-yd class machine very similar in size to the company's previous cancelled L-500 program. The L-600 would make its world debut at the American Mining Congress equipment exhibition, held in Detroit, Michigan, in May 1976. Image taken in the summer of 1978.

The L-600 was offered with two engine choices—a 12-cylinder GM Detroit Diesel 12V-71T, or a 6-cylinder Cummins KT1150, both rated at 525-hp. Pictured is the prototype L600 (S/N 1027) working at a Pennsylvania jobsite on January 4, 1977. Only 26 L-600 LeTro-Loaders were ever produced.

Pictured in mid-1975 is the second prototype L-800 LeTro-Loader (S/N 1028SS) assembled by the company (the first pilot L-800 was actually the rebuilt L-700SS from 1973). The second prototype L-800 would ship to Utah Construction & Mining's Navajo Mine, located near Farmington, New Mexico, in August 1975.

The Marathon LeTourneau L-800 LeTro-Loader was originally offered with the choice of an 860-hp GM Detroit Diesel 16V-92T engine, or an 800-hp Cummins VTA-1710-C800 unit. Bucket capacity was rated at 15 cu-yds, with a 45,000 pound payload limit. Image date Summer 1978.

This L-800 LeTro-Loader (S/N 1180) pictured in May 1981 is equipped with a huge custom-built LeTourneau 55-cu-yd capacity wood-chip bucket. This unit was built for use at a Proctor & Gamble Co. pulp-mill in Grand Prairie, Alberta, Canada. The bucket was just over 18 feet wide and 13 feet deep, with a height of 9 feet, 4 inches.

Shown testing at the Longview plant in 1982 is an L-800 equipped with a one-of-a-kind experimental ejector bucket. This bucket was rated the same as a standard L-800 unit and was designed to be retrofitted to any existing L-800 loader. Total number of all L-800 LeTro-Loaders placed into service was a very healthy 192 machines (not counting the original L-700SS/L-800 prototype).

The wheel-dozer counterpart to the L-800 front-end loader in the Marathon LeTourneau product line was the D-800 LeTro-Dozer. Introduced in 1978, the D-800 was designed primary for heavy earth-moving applications such as land reclamation, or for coal stockpiling. Pictured is the pilot D-800 (S/N 1027) at the Longview plant in the summer of 1978.

The diesel-electric drive D-800 LeTro-Dozer utilized many of the components from the L-800. Unique to the D-800 was its dozer front-end assembly, and its high-visibility operator's cab with expanded front glass viewing areas. Only 18 of the big wheel-dozers were ever built. Pictured is the pilot D-800 operating at Drummond Coal in Kellerman, Alabama, in December 1978.

The smallest front-end loader currently offered by the company is the L-950. Originally introduced in September 2004 as the L-950 "Pit Bull," the loader was equipped with a Switched Reluctance (SR) Propulsion System utilizing traction wheel motors that contained no armatures or brushes. The 18-cu-yd capacity loader has been a slow seller for the company, with most customers interested in this size class of a machine opting for the company's L-1150 instead. Only 25 units had been produced at the time of this writing. Image taken at the JSC Alrosa/Nurbinsky Mine in Russia on February 9, 2010. *Image courtesy LTI.*

The L-950 loader's wheel-dozer counterpart in the product line was the D-950. Introduced at the same time as the L-950, the D-950 "Pit Bull" utilized many of the wheel-loaders components, including its revolutionary SR-drive system. But the mining marketplace can be tough going for a wheel-dozer, and so it was with the D-950. Only nine units were produced before the D-950 was quietly removed from the Joy/P&H product line in early 2014. Pictured is the pilot D-950 testing at the Cortez Gold Mine, located near Crescent Valley, Nevada, on December 1, 2004. *Image by the author.*

Introduced in 1982, the original diesel-electric drive L-1000 LeTro-Loader was classified as a 17-cu-yd capacity machine, with a standard 51,000 pound load rating. Shown is an L-1000 working at Material Services, located in Thornton, Illinois, on July 19, 1996. *Image by the author.*

The L-1000 proved to be a popular model for LeTourneau with a total of 107 units listed as being produced. The last L-1000 rolled off the Longview assembly line in January 2004. This L-1000 is pictured working at ECC Industries in Sylacauga, Alabama, on June 3, 1998. *Image by the author.*

A unique model variation in the L-1000 loader program was the TCL-1000 (Trailing Cable Loader). Officially unveiled at the October 1982 AMC equipment exhibition in Las Vegas, the TCL-1000 was not equipped with a diesel engine. Instead the loader received its power through a rear-mounted transformer set and a retractable electrical cable-drum assembly. Electrical power was supplied by the mine site via a movable cable-boat. Pictured in August 1982 in Longview is the TCL-1000 equipped with a 30-cu-yd coal loading bucket.

The TCL-1000 spent a couple of years working at Utah International's Navajo Mine, located near Farmington, New Mexico, but would never go beyond the prototype testing stage. Pictured at the mine site in April 1984 is the prototype TCL-1000 (S/N 1027) fitted with a redesigned rear cable-drum unit to reduce overall weight and increase reliability.

Another solid performer for the company during the late 1980s and 1990s (both performance- and sales-wise) was the L-1100 LeTro-Loader. Introduced in August 1986, the L-1100 was classified as a 22-cu-yd capacity loader, capable of handling a 66,000-pound bucket payload. This L-1100 is pictured working at the Lee Ranch Mine, located near Grants, New Mexico, on July 10, 1996. *Image by the author.*

The L-1100 looked much like the L-1000, just proportioned larger overall. The quickest way to tell the loaders apart visually was the unique lift-arm design of the L-1100 featuring "Mickey-Mouse" ear shaped reinforcing brackets for the lift-cylinder struts. Image taken on July 10, 1996. *Image by the author.*

Like the L-1000, the larger L-1100 proved very popular in the mining industry, but more so, with 122 units listed as being shipped into service, with the last one leaving the Longview plant in May 2005. Pictured loading a bottom-dump coal hauler on June 16, 2000, was the third L-1100 purchased by the Lee Ranch Mine. *Image by the author.*

One of the newest LeTourneau SR-drive system model designs introduced before Joy Global purchased the company was the L-1150. Officially introduced at the September 2008 MINEXpo show in Las Vegas, the L-1150 is equipped with a standard bucket capacity listed at 25 cu-yds, with a 76,000 pound load rating. By the end of 2013 approximately 16 units had been shipped into service worldwide. Image taken in Longview of the prototype L-1150 on September 5, 2008. *Image courtesy LTI.*

Marathon LeTourneau introduced its L-1200 LeTro-Loader in the fall of 1978. Equipped with a 22-cu-yd bucket rated at 66,000 pounds, it was the largest diesel-electric drive loader model the company had ever produced up until that time. In the marketplace, only the 24-cu-yd Clark Michigan 675C front-end loader was larger. Image date August 1978.

In August 1978, the company shot a series of nighttime publicity photos in Longview, Texas, depicting the L-1200 adorned with Vegas-like showgirls, to be given out in the press packets available at the Las Vegas AMC show held in October 1978. All of the models were former Dallas Cowboy football team cheerleaders that had recently left the squad and formed their own modeling agency. This was their reported first major assignment.

After the Las Vegas AMC show, the pilot L-1200 LeTro-Loader (S/N 1027) was shipped to Fran Contracting, located near Farmington, Pennsylvania, on a lease agreement. Once its contract had expired, it was shipped to Drummond Coal in Alabama (around December 1979). Pictured in December 1978 is the pilot L-1200 at its first working jobsite for Fran Contracting.

It would not be until January 1980 that the second prototype L-1200 LeTro-Loader would ship from the Longview plant to Drummond Coal, in Townley, Alabama. Equipped like the prototype #1 machine, the second loader also featured Goodyear 65.50x51, 54PR (L-5) SXT D&L Nylo-Steel tires. Pictured in February 1980 is the second L-1200 (S/N 1028) built at Drummond Coal.

When the first L-1200 (S/N 1027) produced shipped to Drummond Coal from Fran Contracting, it would be equipped with large Michelin 50.5x51 XRD-2 (L-5) "banana-tread" tires, replacing the original Goodyear's which were not performing up to expectations on a loader the size of the L-1200. Eventually, the first L-1200 was shipped back to the Longview plant, rebuilt, and shipped in August 1982 to C.C.B. in Belgium. Pictured in February 1980 is the first prototype L-1200 (S/N 1027) operating at Drummond Coal, equipped with the big Michelin tires.

The third L-1200 built (which was still considered a prototype unit by LeTourneau) was shipped to Kaiser Resources Ltd., Harmer Ridge, Natal, British Columbia, Canada, in May 1980. This loader utilized several options not to be found on any other L-1200 through its production run. These included a 1,200-hp Cummins KTA2300 engine, United 65.55x51, 60PR-sized tires, and a 36-cu-yd coal loading bucket. Pictured is the third L-1200 (S/N 1029) at Kaiser Resources in July 1980.

Shown in July 1980 is the third L-1200 built (S/N 1029) making good use of its 36-cu-yd coal loading bucket, at Kaiser Resources Ltd., in British Columbia, Canada.

Marathon LeTourneau finally got the L-1200's tire wear problems in hand when they fitted the big loader with the huge Goodyear Deep Tread 67x51, 44PR-sized tires. Pictured in 1983 is one of the last six L-1200 built (all shipped to the El Cerrejon coal mine in Columbia, South America, between 1982 and 1984) equipped with the new 67-inch wide Goodyear tires. The mine retired its last L-1200 loader (S/N 1039, unit #11) on October 22, 1998, with 36,548 hours on the meter. Total number of L-1200 LeTro-Loaders produced by LeTourneau was only 11 machines.

Unveiled in April 1999 at the Longview plant, the L-1350 was the company's first diesel-electric loader design to feature LINCS (LeTourneau Integrated Network Control System), which brought digital speed and precision to the management and monitoring of all the loaders functions. Pictured on June 5, 2000, is the pilot L-1350 loader in operation at a lignite coal mining operation located in northern North Dakota. *Image by the author.*

The L-1350 was the first of the "50" Series loader designs to be released by LeTourneau. Standard bucket fitted to the loader was rated at 28 cu-yds, with an 84,000-pound load limit. Image taken on June 5, 2000. *Image by the author.*

The L-1350 was designed to replace the L-1400 in the company's product line. The original L-1350 was replaced by an updated L-1350 Generation 2 model in May 2011. Total number of L-1350 loaders shipped (of both model types) by the end of 2013 was 59 units. Pictured on May 13, 2002, is an L-1350 (unit #4) operating at Rio Tinto's U. S. Borax mining complex, located near Boron, California. *Image by the author.*

When introduced in 1990, the LeTourneau L-1400 caused quite a buzz in the mining industry. With its standard 28-cu-yd rock bucket, it would lay claim to the title of the "world's largest wheel loader." Pictured in November 1990 is the pilot L-1400 at its first operating jobsite at Geupel's High Powered Energy Coal, located in Drennen, West Virginia.

One of the first problems the L-1400 had to address early on was the availability of suitably-sized tires. Tires originally fitted to the loader were rather skinny-looking 49.5/57, 68PR (L-5) series rubber. Eventually the tire manufactures stepped up to the plate and started to produce larger tires specifically designed for loaders the size of the L-1400 and larger. Shown on October 19, 1995, is an L-1400 (unit #21) in operation at Dry Fork Coal, located just north of Gillette, Wyoming. *Image by the author.*

The L-1400 had a good production run for the company and was a real success in mining operations the world over. Total number of L-1400 loaders produced was 61, with the last unit shipping from Longview in October 2002. Pictured on October 19, 1998, is an L-1400 (unit #40) loading coal at the P&M Kemmerer Mine in Wyoming. *Image by the author.*

The L-1400 would relinquish its crown as the world's largest wheel loader in December 1993, not to a machine produced by a competitive manufacturer, but to another mighty LeTourneau – the L-1800. Pictured on October 19, 1995, is the 33-cu-yd capacity pilot L-1800 that was shipped to the AMAX Eagle Butte Mine, located north of Gillette, Wyoming. *Image by the author.*

The L-1800 was equipped with a standard rock bucket rated at 33 cu-yds, with an overall load capacity of 100,000 pounds. This L-1800 (unit #8) from May 16, 1996, is shown working at Powder River Coal's Rochelle Mine (referred to today as the North Antelope Rochelle Mine, or NARM for short), located near Wright, Wyoming, and is equipped with a special 36-cu-yd combo-bucket rated at 94,000 pounds.

In 1998 the L-1800 working at the Rochelle Mine (now NARM) had its 36-cu-yd bucket replaced with a larger capacity 55-cu-yd coal loading version, the first of its kind ever fabricated by LeTourneau. Total number of L-1800 loaders built was 30 machines, with the last unit leaving the Longview assembly line in July 2002. Pictured is the L-1800 fitted with the 55-cu-yd coal bucket on October 7, 1998. *Image by the author.*

In 2002 the L-1800 was replaced with the L-1850 series featuring the company's LINCS digital monitoring system, and full operator joystick controls. Pictured on May 6, 2002, is the pilot L-1850 working at Peabody Energy's North Antelope Rochelle Mine (NARM), near Wright, Wyoming, equipped with a 55-cu-yd coal loading bucket. This machine originally shipped from the Longview plant in February 2002. *Image by the author.*

The L-1850 carried the same standard 33-cu-yd bucket (100,000 pounds) as the previous L-1800 model, and could be optioned with a 31-cu-yd (94,000 pounds) unit when equipped with high-lift. Shown is the second L-1850 built working at the Cortez Gold Mine, near Crescent Valley, Nevada, on September 21, 2006. It was equipped with a special 33-cu-yd rock-bucket rated at 94,000 pounds and the high-lift loader arm configuration. *Image by the author.*

Since its introduction in 2002, the L-1850 has been the leading seller for the company with machines operating the world over in some of the toughest digging conditions imaginable. In 2011 it was upgraded to a "Generation 2" status (now rated at 40 cu-yds, 120,000 pounds). Total number of original L-1850 loaders produced was 111 units, with an additional 35 "Gen-2" machines ordered by mid-2014. Pictured on December 2, 2004, is an L-1850 (unit #12) operating at the Barrick Goldstrike Mine, near Carlin, Nevada. *Image by the author.*

Holding the title of the "world's largest wheel loader" today is the incredible L-2350. With a standard rock-bucket capacity of 53 cu-yds, rated at 160,000 pounds, no other wheel loader even comes close to the L-2350. The first L-2350 was officially introduced at the October 2000 MIN-Expo in Las Vegas, Nevada. Pictured is an L-2350 (unit #3) operating at the ASAR-CO Ray Mine, located near Hayden, Arizona, on May 30, 2007. *Image by the author.*

Not only does the 586,000 pound L-2350 carry the largest payload of any wheel loader in the world, it is also equipped with the largest tires ever produced for a mobile piece of machinery. The world record tires are Firestone 70/70-57, 82PR SRG (Super Rock Grip) DT (Deep Tread) LD (Loader Dozer) (L-4) series units, each weighing in at 7.37 tons, with an overall width of 74.2 inches (static loaded width is 69 inches), that are approximately 13 feet in diameter. Image taken May 31, 2007. *Image by the author.*

In 2011 the "Generation 2" version of the giant loader was introduced, with the first unit shipping in July of that year. Total number of the first model type of the L-2350 was 17, with an additional 25 "Gen-2" units in the books by mid-2014. Pictured on July 19, 2010, is an L-2350 in operation at the Antamina copper-zinc mine, located in the Andes Mountains of Ancash, Peru. *Image courtesy LTI.*

Marathon LeTourneau would add off-highway mining haulers to its product line in February 1985, after it purchased the diesel-electric drive "Titan" truck line from Diesel Division, General Motors Canada, Ltd. This included all engineering and design data, and all intellectual property and machine tools associated with the hauler line. Pictured on June 1985 is the first Marathon LeTourneau-built 33-15C Titan (S/N 1032) to come off the Longview assembly line.

The largest capacity trucks to be manufactured by Marathon LeTourneau were its T-2240 Titan haulers. The company only built four of these 240-ton units, with two going to Westar in British Columbia, Canada, in 1990, and two to Syncrude in northern Alberta, in 1992. Total number of all Titan haulers produced by the company in Longview, Texas, was 203 (11 of the 33-15Cs; 6 of the 33-15Ds; and 186 of the T-2000s), with the last one shipping from the Longview plant in January 1996. Shown is one of the T-2240 Titans being loaded at Westar's Balmer pit in November 1990.

In August 1986 Marathon LeTourneau changed the designation of its hauler line-up, now referring to them as T-2000 series trucks. The new truck series featured a new operator's cab design, as well as company-built major electrical components replacing previous GM designs. Pictured on March 11, 1988, is a 200-ton capacity T-2200 Titan operating at the BHP Saraji Mine in Queensland, Australia.

Syncrude would eventually sell both of their T-2240 Titan haulers to North American Construction in the late 1990s, which would in turn pass them on to Suncor (located north of Fort McMurray, Alberta, Canada) after several years of operation. Suncor would modify one of the units (S/N 1185) and fit it with a TowHaul towing package (pictured), capable of handling the largest trucks in operation at the mine site. Image taken on October 25, 2013. *Image courtesy Keith Haddock.*

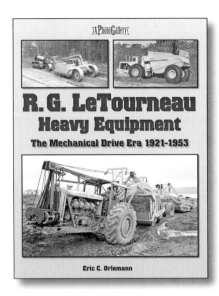
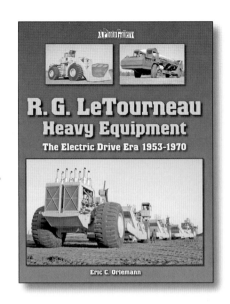
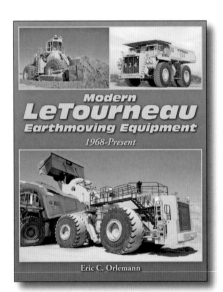
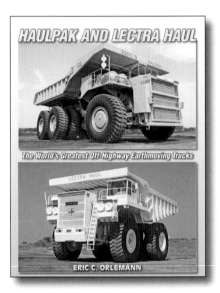
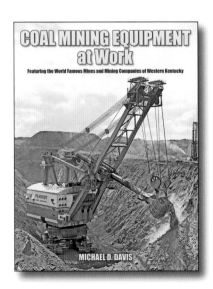

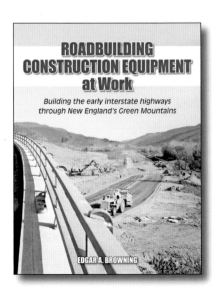